Notes on Fermat's Last Theorem

CANADIAN MATHEMATICAL SOCIETY SERIES OF MONOGRAPHS AND ADVANCED TEXTS

Monographies et Études de la Société Mathématique du Canada

EDITORS: Jonathan M. Borwein and Peter B. Borwein

Frank H. Clarke **Optimization and Nonsmooth Analysis*

Erwin Klein and Anthony C. Thompson **Theory of Correspondences: Including Applications to Mathematical Economics*

I. Gohbert, P. Lancaster, and L. Rodman *Invariant Subspaces of Matrices with Applications*

Jonathan M. Borwein and Peter Borwein *Pi and the AGM A Study in Analytic Number Theory and Computational Complexity*

John H. Berglund, Hugo D. Jünghenn, and Paul Milne **Analysis of Semigroups: Function Spaces Compactifications Representation*

Subhashis Nag *The Complex Analytic Theory of Teichmüller Spaces*

Manfred Kracht and Erwin Kreyszig **Methods of Complex Analysis in Partial Differential Equations with Applications*

Ernest J. Kani and Robert A. Smith *The Collected Papers of Hans Arnold Heilbronn*

Victor P. Snaith **Topological Methods in Galois Representation Theory*

Kalathoor Varadarajan *The Finiteness Obstruction of CTC Wall*

G. Watson **Statistics on Spheres*

F. A. Sherk, P. McMullen, A. Thompson, and A. Weiss *Kaleidoscopes: Selected Writings of H.S.M. Coxeter*

Peter A. Fillmore *A User's Guide to Operator Algebras*

Alf van der Poorten *Notes on Fermat's Last Theorem*

---

*Indicates an out-of-print title.

Notes on Fermat's Last Theorem

ALF VAN DER POORTEN

Centre for Number Theory Research
School of Mathematics, Physics, Computing and Electronics
Macquarie University
Sydney, Australia

A Wiley-Interscience Publication

JOHN WILEY & SONS, INC.

New York • Chichester • Brisbane • Toronto • Singapore

This text is printed on acid-free paper.

Published simultaneously in Canada.

Library of Congress Cataloging-in-Publication Data:

van der Poorten, A. J.
Notes on Fermat's last theorem / Alf van der Poorten.
p. cm.—(Canadian Mathematical Society series of monographs and advanced texts)
"A Wiley-Interscience publication."
Includes bibliographical references and index.
ISBN 0-471-06261-8 (cloth : alk. paper)
1. Fermat's last theorem. I. Title. II. Series.
QA244.V36 1996
512'.74—dc20 95-46319

Printed in the United States of America

10 9 8 7 6 5 4 3

To Joy

requires also the isogeny theorem proved by Faltings (and earlier by Serre when E has nonintegral j-invariant, a case which includes the semistable curves). We note that if E is modular then so is any twist of E, so we could relax condition (i) somewhat.

The important class of semistable curves, i.e., those with square-free conductor, satisfies (i) and (iii) but not necessarily (ii). If (ii) fails then in fact ρ_0 is reducible. Rather surprisingly, Theorem 0.2 can often be applied in this case also by showing that the representation on the 5-division points also occurs for another elliptic curve which Theorem 0.3 has already proved modular. Thus Theorem 0.2 is applied this time with $p = 5$. This argument, which is explained in Chapter 5, is the only part of the paper which really uses deformations of the elliptic curve rather than deformations of the Galois representation. The argument works more generally than in the semistable case but in this setting we obtain the following theorem:

THEOREM 0.4. *Suppose that E is a semistable elliptic curve defined over* $\mathbf{Q}$. *Then E is modular.*

More general families of elliptic curves which are modular are given in Chapter 5.

In 1986, stimulated by an ingenious idea of Frey [Fr], Serre conjectured and Ribet proved (in [Ri1]) a property of the Galois representations associated to modular forms which enabled Ribet to show that Theorem 0.4 implies 'Fermat's Last Theorem'. Frey's suggestion, in the notation of the following theorem, was to show that the (hypothetical) elliptic curve $y^2 = x(x + u^p)(x - v^p)$ could not be modular. Such elliptic curves had already been studied in [He] but without the connection with modular forms. Serre made precise the idea of Frey by proposing a conjecture on modular forms which meant that the representation on the p-division points of this particular elliptic curve, if modular, would be associated to a form of conductor 2. This, by a simple inspection, could not exist. Serre's conjecture was then proved by Ribet in the summer of 1986. However, one still needed to know that the curve in question would have to be modular, and this is accomplished by Theorem 0.4. We have then (finally!):

THEOREM 0.5. *Suppose that $u^p + v^p + w^p = 0$ with $u, v, w \in \mathbf{Q}$ and $p \geq 3$, then $uvw = 0$.*

The second result we prove about the conjecture does not require the assumption that ρ_0 be modular (since it is already known in this case).

Finally

Annals of Mathematics **141** (3), May 1995

Introduction

Fermat's Last Theorem states that there are no positive integers x, y, and z with

$$x^n + y^n = z^n$$

if n is an integer greater than 2. For n equals 2 there are many solutions:

$$3^2 + 4^2 = 5^2, \quad 5^2 + 12^2 = 13^2, \quad 8^2 + 15^2 = 17^2, \quad \ldots$$

the Pythagorean triples.

In the margin of his copy of Diophantus's *Arithmetica* the French jurist Fermat wrote c.1637 that for greater n no such triples can be found; he added that he had a marvelous proof for this, which however, the margin was too small to contain.

There should be no doubt that Fermat was mistaken in thinking that he had a general proof, and that he quickly realized that he was in error. Nonetheless, his claim went into legend as Fermat's Last Theorem and proceeded to be a major motivation in the development of mathematics.

Such was the notoriety of the question that its apparent solution* (as announced in June 1993), as a corollary of the work of Andrew Wiles on the Taniyama-Shimura Conjecture for arithmetic elliptic curves, was a major event in mathematical history.

Fermat's Last Theorem is very important to mathematics, and yet of no importance at all. But its fame is indisputable. I have long been a reader of science fiction. Imagine my delight when I first saw that the devil was thwarted by the FLT in Arthur Porges, 'The Devil and Simon Flagg', *The Magazine of Fantasy and Science Fiction*, New York, 1954. Happily for me, for that's where I now have it,† it was reprinted in Clifton Fadiman, *Fantasia Mathematica* (New York: Simon and Schuster, 1958), pp. 63–69.

It happened that I was scheduled in mid-1993 to begin a series of "extra" lectures to our best undergraduate students. The opportunity that Wiles presented led me to talk about Fermat's Last Theorem and, of course, to expand upon various pieces of history and elementary mathematics that

*What with various alarums, I had to keep changing this footnote; many of these lectures were written while the status of the proof was still in some doubt. In late 1994 I had the additional remark "There is now general agreement that the public preprints Andrew Wiles, 'Modular elliptic curves and Fermat's Last Theorem' and Andrew Wiles with Richard Taylor, 'Ring-theoretic properties of certain Hecke algebras' of late October 1994, contain a proof." Now I can add that the papers have appeared in *Annals of Mathematics*.

† *You* now have it reprinted as an appendix to this book.

seemed apropos. Sadly, all too many students are barely aware that it is possible to read mathematics and about mathematics, and so can have little idea either of the history of Fermat's Last Theorem or of the content and spirit of the work that leads to its solution. I remembered having given some lectures on the FLT at various meetings some 15 years ago, so first that stuff was dusted off and TEXed. As I progressively wrote the notes, and gave lectures introducing them, I realized that it might not be as impossible as I had first thought to give a reasonable inkling of the ideas that underlie the eventually successful arguments.

One of the difficulties in reading, or listening to, mature mathematics is its immense vocabulary and the volume of notions that seems to be required. Nor can one readily discover the meaning of the more popular ideas because all too often they are defined in terms of yet more obscure words. The truth is, fortunately, that few — perhaps none — of us know all the definitions. We rely on a feeling for what must be intended, knowing that we can refine that feeling should needs be. In a sense, these notes should be seen precisely as an attempt to create some useful feelings.

The style I have adopted in the notes is to *announce* all sorts of things. Some announcements are just definitions, others are facts whose explanations we are not yet in a position to comprehend. But many of my claims are indeed obvious after one has thought a little while. Mostly, these accessible claims are signaled by such phrases as "we see that", "it is now obvious that", "clearly", and the like. Throughout, the exercises for the reader are to supply the extra remark needed to make my claims totally obvious or to explain why my hints really make my claim immediate.

I indulge myself with footnotes mentioning "facts" that amuse me, and with the odd anecdote. Those stories are the kind we tell one another in the evenings after a heavy conference day and with which we bore our partners.

The first few lectures barely have any prerequisites at all beyond a rather good high school background in mathematics. But generally, the obvious prerequisite for some understanding of what I write is a tolerance for and interest in formulas. After a while, I also suppose acquaintance with a first course in linear algebra. Mind you, rather more than that is required for the details.

I do not attempt to replace a library of textbooks and learned articles. I am vague and I jump all over the place. But my remarks should contain sufficiently many understandable oases to remain interesting. Just don't be balked by the intervening, difficult to understand, deserts. And then you and I had better go off elsewhere to do the hard work necessary to acquire all the details we may find necessary to be fully satisfied. In the meantime I have learned a great many things that I barely knew I knew. Similarly you, the reader, will have picked up a few notions and ideas.

I commence with quasi-historical remarks to set the scene. For some of the real history one should first go to Harold M. Edwards, *Fermat's Last Theorem*, Graduate Texts in Mathematics **50** (New York: Springer-Verlag, 1977) and also to his historical articles; then Paulo Ribenboim's *13 Lectures on Fermat's Last Theorem* (New York: Springer-Verlag, 1979) provides a plethora of detail. For the legends one turns to Eric Temple Bell's *The Last Theorem*, (Washington, D.C.: Mathematical Association of America, 1990). For the historical context it would be hard to excel André Weil, *Number Theory: An approach through history. From Hammurapi to Legendre*, (Basel, Switzerland: Birkhäuser 1984). Then I proceed to mention all sorts of odds and ends in an effort to sneak up on Wiles' argument without becoming too tangled in incomprehensible detail.

The point is to glimpse all sorts of exciting pieces of mathematics and to be moved to teach ourselves more. Among my motives in giving these lectures was that of trying to make mathematics a little less boring. All too often the reason for the incomprehensible things one is asked to learn is "beyond the scope of the text". That seems a constipated approach to me. It's not the way mathematicians talk to one another in seminars, colloquia, or in our offices and corridors. We wave our arms and try to pinpoint the critical thoughts. My idea was to try to provide motive — and damn the details. Mind you, lots of things are not said. I confined myself religiously to four typeset TEX pages per lecture, changing my wording (not the facts, I hope) to fit in. That restriction makes less sense now that these "Notes" appear in book form, but I've lapsed from the principle only to the extent of adding extensive remarks and comments, sometimes exactly of the kind I made in the actual lectures.

When an undergraduate, it didn't seem to bother me all that much that I barely understood the mathematics which I could so carefully reproduce in examinations. I firmly maintained the viewpoint that mathematics is a branch of magic. If one knows the *true names* of things, one can control them; of course, I mean knowing definitions and using a sensible notation. Just so, casting the right *spell* at the correct time is announcing the appropriate theorem in suitable wording. Casting such spells would induce my teachers to believe that I understood lots of mathematics and would cause them to give me high marks, scholarships, and jobs. My viewpoint worked. One day it dawned upon me that I wasn't just copying spells but was constructing my own; and pieces of the unruly mathematical world had been tamed by the true names that I had given them.

And what should you, the reader, know? I expect you to want to know about Fermat's Last Theorem and to be willing to pick up a few bits of mathematics on the way. Or better, I expect you to want to pick up a few bits of mathematics and not to object to my talk about Fermat's Last Theorem. I hope you like puzzling through and checking complicated identities. I do assume a tolerance for ambiguity: the will to press on when doubtful about what's just been said (and that includes being certain that one doesn't understand some section). The Australian injunction: "Not to worry. She'll be right, mate", should be on your lips at all times. So I'll gaily use and misuse words, both technical and ordinary, whose meaning you may well not know, and I invite you not to worry. If it's at all important in the sequel, then you'll probably develop at least a feeling for the meaning, and if it's not important, well then, it's of no importance.

I was rather slower in finishing writing these notes than I should have been, so by the time I was thinking of concluding them (October 1994) it had become quite obvious that the gap in Wiles' arguments was serious. Fermat's Last Theorem had not yet been fully settled. Fortunately, such a mishap is very much in the spirit of these notes. It was anyhow generally agreed that Wiles' arguments had shown the way that the Last Theorem would eventually be settled. In any case it wasn't as if the detail I was about to present would be relevant to the lacuna in the proof.

I had barely written the preceding sentences before Wiles released a pair of papers, his 'Modular elliptic curves and Fermat's Last Theorem' and an addendum jointly with Richard Taylor 'Ring-theoretic properties of certain Hecke algebras', which settle the matter after all (this occurred on October 25, 1994). Although the various lectures are not marked with the date I wrote or gave them, it will probably be evident from the various asides just when in the period August 1993–August 1995 they were concluded. Perhaps this "immediacy" will strengthen the reminder I want to give that mathematics is very much a living thing, not one cut and dried in some past century.

There's been an unusual amount of publicity for mathematics arising from Wiles' work, enough to convince one, if one needed the convincing, that proving Fermat's Last Theorem is important. Mind you, a great deal of that publicity was quite inane. My 'Remarks on Fermat's Last Theorem', which follows the Lectures forming the bulk of this book, were an attempt by me to write something that might have been of interest to a serious newspaper. I did not find one and eventually, lest they blush unseen, published the 'Remarks' in the *Australian Mathematical Society Gazette*.

The proof of Fermat's Last Theorem is important. A reasonable comparison is to suggest that it's as important to mathematics as the landing on the moon was to science and technology. The proof is as dramatic as the landing, and as exciting to the onlookers. Of itself, knowing that Fermat's claim is true barely advances mathematics, but then the actual taking of people to the moon did not hugely add to our scientific knowledge. The moon landing was the corollary, as it were, to a long and steady scientific and technological advance, punctuated by occasional dramatic breakthroughs. The landing was not itself such a breakthrough. In that spirit, Fermat's Last Theorem is a culmination of some 350 years, well certainly 250 years, of mathematical advance. However, Wiles' work does constitute a dramatic advance, one of those special watersheds. The great advance lies in his showing that, indeed, the Modularity Conjecture of Taniyama–Shimura–Weil is true, at any rate for semistable elliptic curves. This confirms that the experimental and contextual evidence had not led mathematics astray. If the "Holy Grail" of Fermat's Last Theorem was needed to motivate that advance, well that's fine. There will have been worse reasons for advancing mathematics.

I shall not try to list the mathematicians and other interested people who have aided me in preparing the following pages. Were it not for their help there would be many more errors left uncorrected. I also leave anonymous, here, the various 'poets', who failed to show decent shame and seemed eager for their work to be promulgated. I do acknowledge the help of Catherine Goldstein, C. J. Mozzochi, Kate van der Poorten, and Andrew Wiles in making available the pictures that decorate this book.

I particularly want to mention my indebtedness to my colleague Ross Moore, who implemented most of the technical layout by combining and customizing various LaTeX packages, as well as doing some preliminary editing. The elegant mathematical diagrams which appear throughout the book were created by him using *Mathematica**, with labels and annotations added using XY-pic†, of which Ross is one of the authors. The eps-picture "Rational points on the elliptic curve $x^3 + y^3 = 9$" on page 59 appears in the XY-pic Reference Manual, Version 3.x (Copyleft: Kristoffer Rose and Ross Moore) available electronically on the Web/Internet. It is used there as an example illustrating the ability to import graphics and easily position extra text/labeling around and over the graphic, using its intrinsic coordinate system.

I had better, moreover, apologize to my wife Joy, and my children Kate and David, for having used my work on this book seemingly as an excuse to hide from them. Certainly, that was the opinion of our dog, Talleyrand. I am delighted to dedicate this book to Joy.

I am grateful to the American Mathematical Society's gopher server, which carried a preliminary version of the first eight lectures and thereby, by way of David Cox as intermediary, interested Wiley-Interscience in publishing these notes.

Macquarie University, Sydney, Australia
November, 1995

Work on this book was supported in part by grants from the Australian Research Council and a research agreement with Digital Equipment Corporation.

* *Mathematica*, a registered trademark of Wolfram Research Inc., is software both for doing mathematics and for constructing graphic visualizations and presentations.

† XY-pic is free software for typesetting mathematical diagrams in TeX and LaTeX; browse the Web pages at `http://www.diku.dk/diku/users/kris/Xy-pic.html` and `http://www.mpce.mq.edu.au/~ross/Xy-pic.html`.

The PROMYS Shirt

FERMAT'S LAST THEOREM: *Let $n, a, b, c \in \mathbb{Z}$ with $n > 2$. If $a^n + b^n = c^n$ then $abc = 0$.*

Proof: The proof follows a program formulated around 1985 by Frey and Serre [**F, S**]. By classical results of Fermat, Euler, Dirichlet, Legendre, and Lamé, we may assume that $n = p$, an odd prime ≥ 11. Suppose $a, b, c \in \mathbb{Z}$, $abc \neq 0$, and $a^p + b^p = c^p$. Without loss of generality we may assume that $2 \mid a$ and $b \equiv 1 \bmod 4$. Frey [**F**] observed that the elliptic curve $E : y^2 = x(x - a^p)(x + b^p)$ has the following "remarkable" properties: (1) E is semistable with conductor $N_E = \prod_{l \mid abc} l$; and (2) $\overline{\rho}_{E,p}$ is unramified outside $2p$ and is flat at p. By the modularity theorem of Wiles and Taylor-Wiles [**W, T-W**], there is an eigenform $f \in S_2(\Gamma_0(N_E))$ such that $\rho_{f,p} = \rho_{E,p}$. A theorem of Mazur implies that $\overline{\rho}_{E,p}$ is irreducible, so Ribet's theorem [**R**] produces a Hecke eigenform $g \in S_2(\Gamma_0(2))$ such that $\rho_{g,p} \equiv \rho_{f,p} \bmod \mathfrak{p}$ for some $\mathfrak{p} \mid p$. But $X_0(2)$ has genus zero, so $S_2(\Gamma_0(2)) = 0$. This is a contradiction and Fermat's Last Theorem follows. Q.E.D.

References

[**F**] Frey, G.: Links between stable elliptic curves and certain Diophantine equations. *Ann. Univ. Sarav.* **1** (1986), 1-40.

[**R**] Ribet, K.: On modular representations of Gal($\overline{\mathbb{Q}}/\mathbb{Q}$) arising from modular forms. *Invent. Math.* **100** (1990), 431-476.

[**S**] Serre, J.-P.: Sur les représentations modulaires de degré 2 de Gal($\overline{\mathbb{Q}}/\mathbb{Q}$). *Duke Math. J.* **54** (1987), 179-230.

[**T-W**] Taylor, R. L., Wiles, A.: Ring-theoretic properties of certain Hecke algebras. *Annals of Math.* **141** (1995), 553-572.

[**W**] Wiles, A.: Modular elliptic curves and Fermat's Last Theorem. *Annals of Math.* **141** (1995), 443-551.

It doesn't fit the margin,
*But it does go on a shirt.**

*Couplet from a 'poem' of Fernando Gouvêa remarking on the information printed on the T-shirts sold at the Boston University meeting on *Fermat's Theorem*, August, 1995. The references appear on the back of the shirt. PROMYS is a mathematics program for bright and ambitious high school students. The program is directed by Glenn Stevens and sponsored by Boston University with financial support from the National Science Foundation. The T-shirt was designed by members of the 1995 PROMYS counselor staff who attended the Boston University Fermat Conference.

Biographical Information

Alfred Jacobus van der Poorten was born in Amsterdam in 1942 and spent the war as "Fritsje Teerink", believing himself to be the youngest child of a family in Amersfoort. His true parents, David and Marian, were among the few who returned from the camps. His family migrated to Sydney, Australia when he was eight years old. He quickly learned to barrack for St George* and, a little later, to support Carlton†. Were it not for [the] commonsense [of his wife Joy], he would quite probably now be a politician.

Alf survived Sydney Boys' High School and, in accepting the bribe of a university cadetship, chose to study mathematics at the University of New South Wales, in Sydney, obtaining a B.Sc. with first class honours and a Ph.D. degree. He avoided the compulsory general studies subjects of the science degree by pursuing a concurrent major sequence in philosophy, which he later converted to a B.A. with honours. Lest his education be totally impractical, he then proceeded to complete an M.B.A.degree. Alf verifies the definition that "a professor is a person who goes to university and never comes out"; he did not leave UNSW until he took up a chair of mathematics at Macquarie University, in 1979.

During the sixties, Alf was active in student politics. He was president of the Student's Union Council 1964-65 and president of the University Union 1965-1967. Alf was declared "Young 'refo' of the year" in 1966 [he received the Australian Youth Citizenship Award "for his attainments in community service, academic achievement and youth leadership whereby he has set an outstanding example to the community"]. He represented the undergraduates on the Council of the University [its "Board of Governors"] 1967-1969 and represented the staff [faculty] unions 1969-1973. He was a member of the federal executive of the Federation of Australian University Staff Associations in the early seventies.

Alf was appointed Lecturer in Mathematics in 1969, promoted to Senior Lecturer 1972 and, after a year's leave spent at the Rijksuniversiteit Leiden and at Cambridge, to Associate Professor in 1976. Then in 1979, after a sabbatical leave at Queen's University, Kingston, Ontario, he moved to

*The St George Rugby League Football Club benefited from his dedicated attendance at 74 consecutive matches 1954-1957 by proceeding to win the premiership for eleven consecutive years 1956-1966.

†The "Mighty Blues" hale from Sydney's rival city Melbourne; their game is Australian Rules Football. They won the flag in 1995.

Macquarie University as Professor of Mathematics. He was elected Head of the School of Mathematics and Physics and served as Head of School 1980–1987. In the meantime he spent periods at the Université Bordeaux I, at TH Delft, and at MSRI, Berkeley. For two years, 1986–87, he was Vice-President of Academic Senate (Chairman of the Academic Board). Alf was again elected as Head, now of the renamed School of Mathematics, Physics, Computing and Electronics, in April, 1991. Alf has been a thorn in the side of successive university administrations. That's been exacerbated by his being elected to represent the academic staff on the Council of Macquarie University 1986–87, and since 1989.

Alf spends a lot of time playing with his Macintosh computer. Luckily, he can do that simultaneously with watching football on TV [He's quite unprejudiced and can get equal pleasure from rugby league, Australian rules, American football, soccer or rugby union]. Then there's cricket, baseball, ... In the interstices he reads science fiction and mysteries. He claims never to have thrown a book away and thus owns some five thousand science fiction books and several thousand mysteries; but he's not really a collector, just a keeper. Naturally, this book was written on a Macintosh, using TeX as implemented in Textures by Blue Sky Research.

In real life, Alf is a dedicated research mathematician. He has written some 120 papers and enjoys adding to his United Airlines frequent flyer miles by attending conferences as if there were no tyranny of distance. His current research activity includes work on 'Continued fractions' and 'Effective diophantine approximation'. Alf is currently a member of the Council of the Australian Mathematical Society. In 1994–95, Alf chaired a Working Party on behalf of the National Committee for Mathematics to report to the Australian Research Council on "Mathematical Sciences Research and Advanced Mathematical Services in Australia".

Contents

Appendices

Notes on Fermat's Last Theorem

Notes on Fermat's Last Theorem

LECTURE I

The story of "Fermat's Last Theorem"
has been told so often it hardly bears retelling.
H. M. Edwards

Dramatis Personæ:

Euclid of Alexandria	~ -300
Diophantus of Alexandria	$\sim$ 250
Pierre de Fermat	1601-1665
Leonhard Euler	1707-1783
Joseph Louis Lagrange	1736-1813
Sophie Germain	1776-1831
Carl Friedrich Gauss	1777-1855
Augustin Louis Cauchy	1789-1857
Gabriel Lamé	1795-1870
Peter Gustav Lejeune Dirichlet	1805-1859
Joseph Liouville	1809-1882
Ernst Eduard Kummer	1810-1893
Harry Schultz Vandiver	1882-1973

Gerhard Frey
Kenneth A. Ribet
Andrew J. Wiles

Fermat's Last Theorem states that there are no positive integers x, y, and z with

$$x^n + y^n = z^n$$

if n is an integer greater than 2. For n equals 2 there are many solutions:

$$3^2 + 4^2 = 5^2, \quad 5^2 + 12^2 = 13^2, \quad 8^2 + 15^2 = 17^2, \quad \ldots$$

the Pythagorean triples. In the margin of his copy of the *Arithmetica* of Diophantus the French jurist Fermat wrote *circa* 1637 that for greater n no such triples can be found; he added that he had a marvelous proof for this, which however the margin was too small to contain:

> *Cubum autem in duos cubos, aut quadrato-quadratum in duos quadrato-quadratos, et generaliter nullam in infinitum ultra quadratum potestatem in duos ejusdem nominis fas est dividere; cujus rei demonstrationem mirabilem sane detexi. Hanc marginis exiguitas non caperet.*

Every other result which Fermat had announced in like manner had long ago been dealt with; only this one, the *last*, remained.

QVÆSTIO VIII.

PRopositvm quadratum diuidere in duos quadratos. Imperatum sit vt 16. diuidatur in duos quadratos. Ponatur primus 1 Q. Oportet igitur 16 — 1 Q. æquales esse quadrato. Fingo quadratum a numeris quotquot libuerit, cum defectu tot vnitatum quod continet latus ipsius 16. esto a 2 N. — 4. ipse igitur quadratus erit 4 Q. + 16. — 16 N. hæc æquabuntur vnitatibus 16 — 1 Q. Communis adiiciatur vtrimque defectus, & a similibus auferantur similia, fient 5 Q. æquales 16 N. & fit 1 N. $\frac{16}{5}$ Erit igitur alter quadratorum $\frac{256}{25}$, alter verò $\frac{144}{25}$ & vtriusque summa est $\frac{400}{25}$ seu 16. & vterque quadratus est.

ΤΟΝ ἐπιταχθέντα τετράγωνον διελεῖν εἰς δύο τετραγώνους. ἐπιτετάχθω δὴ τὸν ις διελεῖν εἰς δύο τετραγώνους. καὶ τετάχθω ὁ πρῶτος δυνάμεως μιᾶς. δεήσει ἄρα μονάδας ις λείψει δυνάμεως μιᾶς ἴσας εἶναι τετραγώνῳ. πλάσσω τὸν τετράγωνον ἀπὸ ςς. ὅσων δή ποτε λείψει τοσούτων μο ὅσων ἐστὶν ἡ τῶν ις μο πλευρά. ἔστω ςς β λείψει μο δ. αὐτὸς ἄρα ὁ τετράγωνος ἔσται δυνάμεων δ μο ις λείψει ςς ις. ταῦτα ἴσα μονάσι ις λείψει δυνάμεως μιᾶς. κοινὴ προσκείσθω ἡ λεῖψις, καὶ ἀπὸ ὁμοίων ὅμοια. δυνάμεις ἄρα ε ἴσαι ἀριθμοῖς ις. καὶ γίνεται ὁ ἀριθμὸς ις πέμπτων. ἔσται ὁ μὲν σνς εἰκοστοπέμπτων. ὁ δὲ ρμδ εἰκοστοπέμπτων, καὶ οἱ δύο συντεθέντες ποιοῦσι υ εἰκοστόπεμπτα, ἤτοι μονάδας ις. καὶ ἔστιν ἑκάτερος τετράγωνος.

OBSERVATIO DOMINI PETRI DE FERMAT.

CVbum autem in duos cubos, aut quadratoquadratum in duos quadratoquadratos & generaliter nullam in infinitum vltra quadratum potestatem in duos eiusdem nominis fas est diuidere cuius rei demonstrationem mirabilem sane detexi. Hanc marginis exiguitas non caperet.

Problem 8 in Book II of Claude Bachet's translation of Diophantus asks for a rule for writing a square as the sum of two squares. The resulting equation $z^2 = y^2 + x^2$ is that of the Theorem of Pythagoras, which says that in every right-angled triangle the square on the hypotenuse is the sum of the squares on the other two sides. The logo of Macquarie University's ceNTRe for Number Theory Research

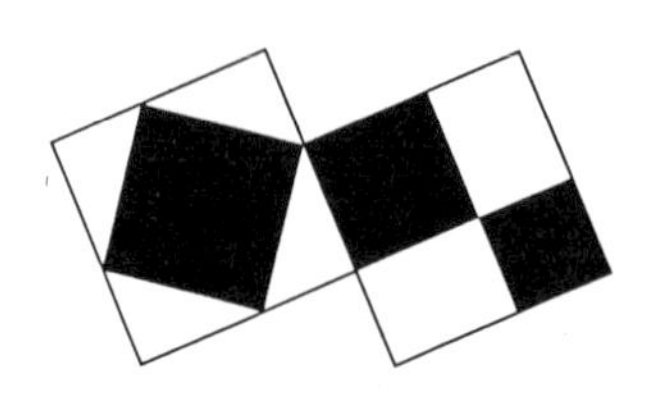

provides a graphical proof; we see that Pythagoras's Theorem has only the depth of the familiar quadratic identity $(x+y)^2 = x^2 + 2xy + y^2$.

It is a little more difficult to find all solutions in integers, but not much more. If $x^2 + y^2 = z^2$, we can suppose that x, y, and z pairwise have no common factor, for such a factor would be common to all three quantities and can be factored out, leaving an equation of the original shape. Thus at least two of x, y and z must be odd. But the square of an odd number, so of the shape $(2m+1)^2 = 4m^2 + 4m + 1$, leaves a remainder of 1 on division by 4 (and on division by 8), while the square of an even number, so of the shape $(2m)^2 = 4m^2$, leaves a remainder of 0 on division by 4. It follows that z must be odd and that one of x and y, say $x = 2x'$, must be even. Then we obtain

$$4x'^2 = z^2 - y^2 = (z+y)(z-y) \quad \text{so} \quad x'^2 = \tfrac{1}{2}(z+y)\tfrac{1}{2}(z-y)\,.$$

But if the product of two numbers that have no factor in common is a square, then each of the two numbers is a square.

This is clear on splitting the two numbers into their prime factors and checking the contribution of each distinct prime. To apply the principle we need only note that both $\frac{1}{2}(z+y)$ and $\frac{1}{2}(z-y)$ are integers, because both z and y are odd; and that they have no common factor. The latter is evident, because if d were a common factor, then d is a factor both of their sum z, and their difference y. Yet we began by determining that y and z are *relatively prime* — that they have no common factor. So both $\frac{1}{2}(z+y)$ and $\frac{1}{2}(z-y)$ are squares, say

$$\tfrac{1}{2}(z+y) = u^2 \quad \text{and} \quad \tfrac{1}{2}(z-y) = v^2\,.$$

Thus $x'^2 = u^2v^2$. Summarizing, we have

$$x = 2uv\,, \quad y = u^2 - v^2\,, \quad \text{and} \quad z = u^2 + v^2\,.$$

We obtain all Pythagorean triples without common factor by choosing integers u and v without common factor and of different *parity* — that is, one odd and the other even, and with u greater than v.

Of course, it is easy to verify that indeed

$$(2uv)^2 + (u^2 - v^2)^2 = (u^2 + v^2)^2\,.$$

Fermat did show that the equation

$$x^4 + y^4 = z^4$$

has no solution in positive integers. In fact, he shows a little more, that already

$$x^4 + y^4 = w^2$$

has no solution in positive integers. As above, we may suppose that x is even, and y and w are odd. Then by the preceding argument, it follows that there are integers a and b so that

$$x^2 = 2ab\,, \quad y^2 = a^2 - b^2\,, \quad \text{and} \quad w = a^2 + b^2\,.$$

From the expression for y^2 it follows that a is odd and b is even, and from that for x^2 we may deduce that there are integers c and d so that

$$a = c^2 \quad \text{and} \quad b = 2d^2 .$$

Hence

$$y^2 = c^4 - 4d^4 .$$

Again applying the results from the case $n = 2$ we see that there are integers e and f so that

$$y = e^2 - f^2 , \quad d^2 = ef \quad \text{and} \quad c^2 = e^2 + f^2 .$$

Clearly, e and f must be relatively prime, so there are integers u and v such that

$$e = u^2 , \quad f = v^2 , \quad \text{and} \quad u^4 + v^4 = c^2 .$$

But now Fermat makes a truly marvelous observation. He notes that c is less than w. So what this argument shows is that given a solution (x, y, w) there is a *smaller* solution (u, v, c) ! That is, eventually, absurd. By the *method of infinite descent*, here introduced, it follows that there is no solution in positive integers to $x^4 + y^4 = w^2$, and *a fortiori* — all the more so, none for $x^4 + y^4 = z^4$.

It was many years later, in 1753, that Euler dealt with the case $n = 3$. There was an alleged error in the argument, later dealt with by Gauss. Dirichlet and Legendre proved the case $n = 5$ in 1825 and Lamé settled the case $n = 7$ in 1839; Dirichlet had proved the case $n = 14$ in 1832.

On 1 March, 1847, Lamé informed the Parisian *Académie des Sciences* that he had settled the general case. Lamé attributed the basic idea of his proof to Liouville. The idea consisted of working with numbers of the shape

$$a_0 + a_1\zeta + a_2\zeta^2 + \cdots + a_{n-1}\zeta^{n-1} ,$$

where $a_0, a_1, \ldots, a_{n-1}$ are integers and ζ is a complex number with the property that $\zeta^n = 1$, but $\zeta \neq 1$. Here Lamé assumed n to be an odd prime number. It had been known for some time that this assumption does not, of course, impose any restriction in the proof of Fermat's Last Theorem.

With the aid of these numbers, $x^n + y^n$ may be split into n factors:

$$(x + y)(x + \zeta y)(x + \zeta^2 y) \cdots (x + \zeta^{n-1} y)$$

and Fermat's equation then assumes the shape

$$(x + y)(x + \zeta y)(x + \zeta^2 y) \cdots (x + \zeta^{n-1} y) = z^n .$$

To this Lamé applies a generalization of the principle already described in the case $n = 2$, whereby if a product of numbers without common factor is an nth power, then each is an nth power. Lamé assumes that this principle holds for the *cyclotomic* integers he has just introduced and proceeds with an argument showing necessarily one of x or y to be zero.

After Lamé, Liouville addressed the meeting. He pointed out that the idea of using complex numbers was nothing new; one could already meet such numbers in the work of Euler, Lagrange, Gauss, and Jacobi. Moreover, it seemed to him, said Liouville, that Lamé implicitly assumed that unique factorization into primes also held for cyclotomic integers.

A second difficulty is numbers that divide 1, or *units* as we now call them. There is a problem, in that, for example, $-4 \times -9 = 36$, with 36 a square and -4 and -9 relatively prime, while neither is a square. In the cyclotomic case one can see readily that there are many more units than just ± 1. For example, $\zeta + \zeta^{n-1}$ is a unit whenever $n > 1$ is odd. Properties of divisibility by $\zeta + \zeta^{n-1}$ play an important role in Lamé's argument. But $\zeta + \zeta^{n-1}$ divides 1, and therefore every number.

N'y a-t-il pas là une lacune à remplir?
J. Liouville

Some of this material has been liberally borrowed from the introduction to the thesis of Hendrik Lenstra, Jr., 'Euclidean number fields', *Math. Intelligencer* **2** (1980), pp. 6–15, 73–77, 99–103; that work is Copyright © 1980 Springer-Verlag New York, Inc. and is partly reprinted here with permission. The rest comes from things I just knew and from notes of mine of some 17 years ago which were probably much aided by thoughts learned from W. J. LeVeque, *Topics in Number Theory*, (Reading, Mass.: Addison-Wesley 1961), Vol 2.

Notes and Remarks

I.1 The style I have adopted here is to *announce* all sorts of things. Some announcements are just definitions, others are facts whose explanations we are not yet in a position to comprehend. Many of my claims are indeed obvious after one has thought just a little while. Mostly, these accessible claims are signalled by such phrases as "we see that", "it is now obvious that", "clearly", and the like. Throughout, the exercises for the reader are to supply the extra remark needed to make my claims totally obvious, or to explain why my hints really make my claim immediate. But just in case there is not enough to do, let me add a few remarks.

I.2 In that notorious margin, Fermat writes that to split a cube into two cubes, or a fourth (biquadratic) power into two fourth powers, or indeed any higher power unto infinity into two like powers, is impossible, and that he has a marvelous proof for this. But the margin is too narrow to contain it.

His failure to provide a proof is not significant. Most of what we know of Fermat's work derives from the challenges he puts to his correspondents; by and large we can reconstruct the arguments from detailed hints. The Last Theorem, however, does not remain the subject of such challenges and it seems plain* that Fermat quickly realizes that he has an argument at most in the case $n = 4$, and perhaps $n = 3$.

I.3 The appearance in 1621 of the translation into Latin of the then extant books of Diophantus signals the beginning of modern number theory. We owe the infamous Last Theorem to Fermat's son Samuel, who in 1670 reprinted his father's copy of Bachet's Diophantus, together with the marginal notes.

I.4 The Macquarie University Number Theory Reports explain on their inside cover that

> The logo for ceNTRe depicts an elegant proof of the theorem of Pythagoras, which states that the area of the square drawn on the hypotenuse of a right angled triangle is equal to the sum of the areas of the squares drawn on the other sides. This result was known to the ancient Babylonians who used it to construct accurate right angles. In the logo, reproduced below with two additional construction lines, the two larger squares each contain four identical right angled triangles, and have sides of the same length, namely the sum of the lengths of the two shorter sides of the right angled triangles. Thus the remaining area in the two squares are equal. In the left hand square, this area is the area of the square drawn on the hypotenuse of the right angled triangle. In the right hand square it is the sum of the areas of the squares drawn on the other two sides.

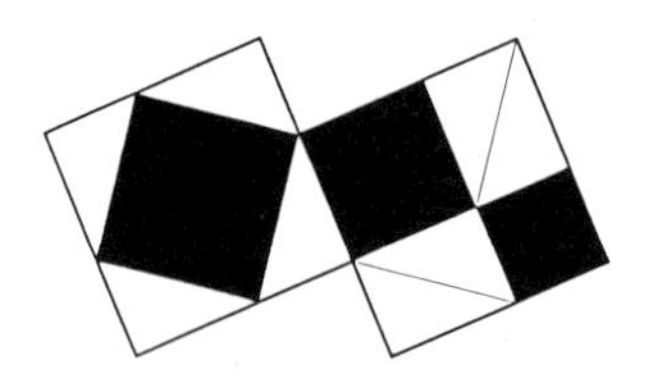

> The sides of the three solid squares in the logo are in the ratio 3:4:5, corresponding to the well known right angled triangle. It also corresponds to the simplest non-trivial solution in integers of the equation $X^2 + Y^2 = Z^2$. Integer solutions of equations such as these play an important role in modern number theory.

---

*Weil, in his *Number Theory: An approach through history. From Hammurapi to Legendre*, (Basel, Switzerland: Birkhäuser 1984), remarks at p104 that "for a brief moment he [Fermat] must have deluded himself into thinking that he had the principle of a general proof; what he had in mind on that day can never be known."

I.5 I really ought to say a little more about descent here, but I'm going to leave that till later. The instructive example I have in mind concerns the fact that a prime number p congruent to 1 modulo 4 can be represented as $p = a^2 + b^2$ — as a sum of two squares. By the way, we see that for such a sum to be prime then, other than $1^2 + 1^2$, it must be 1 modulo 4.

The trick turns out to be that Fermat knows, as will we after my third lecture, that some positive multiple, say kp, of p can be expressed as a sum of two squares. Now Fermat descends on k, showing that also a smaller multiple of p is a sum of two squares. This time, of course, one finishes with the smallest nontrivial multiple $1 \cdot p = p$ and the claim is proved, by descent.

I.6 How about we "prove" Fermat's Last Theorem as many an amateur might. The mistake will be to suppose that if there is a solution then there must be a solution in polynomials $x(t)$, $y(t)$, $z(t)$. This is not a totally ridiculous error. After all, for $n = 2$ there is such a solution $x(t) = 2t$, $y(t) = 1 - t^2$, $z(t) = 1 + t^2$.

In the spirit of Lamé's remark we have $x^n + y^n = z^n = \prod(x + \zeta y)$, and it does follow that each polynomial $(x + \zeta_i y)$ is an n th power w_i^n of some polynomial w_i defined over the complex numbers. But

$$(x + \zeta^2 y) + \zeta(x + y) = (1 + \zeta)(x + \zeta y)$$

yields $a^n + b^n = c^n$ in polynomials $a(t) = w_2(t)$, $b(t) = \sqrt[n]{\zeta}\, b(t)$, and $c(t) = \sqrt[n]{1 + \zeta}\, z(t)$. This is a new solution in polynomials of smaller degree proving, by descent, that there is no such solution. Note where we have used $n > 2$.

I.7 Perhaps it is worth adding the reminder that descent and induction are one and the same thing. The underlying idea is the *Well-Ordering Principle*, which points out the surely obvious fact that any nonempty set of positive integers has a least element. Consider then, for example, the set of positive integers w so that w^2 is the sum of two nonzero integer fourth powers. If this set is nonempty, it has a least element. But the descent argument shows that it does not. Hence the set of such w is empty. Conversely, consider some proposition $P(n)$ for which $P(1)$ is true and with the property that whenever $P(n)$ is true, then also $P(n+1)$ holds. Now turn to the set of integers K for which $P(k)$ is false if k is in K. If K is nonempty, then it has a least element $m + 1$, say, with m positive because $P(1)$ is true. Since $m < m + 1$, we must have $P(m)$ true. Hence $P(m+1)$ is true, contradicting the existence of the minimal element $m+1$ of K. So K is empty and $P(n)$ is true for all positive integers n.

I.8 Additionally, I will eventually explain that the point of studying the equation $x^4 + y^4 = w^2$ in place of $x^4 + y^4 = z^4$ is that the former equation provides an *elliptic curve* to which the method of descent applies. Since also $x^3 + y^3 = z^3$ yields an elliptic curve, it is not unreasonable to suppose that Fermat could handle it as well. But the remaining cases of Fermat's

Last Theorem are geometrically quite different and there is no serious reason whatsoever to think that Fermat had any techniques to attack such equations.

I.9 It surely is obvious that to prove Fermat's Last Theorem it is enough to deal just with the cases $n = 4$ and exponent n prime. For if n is composite, then it is divisible either by some odd prime p or by 4. And for example, an n th power of an integer x is a p th power of $x^{n/p}$. This is an integer if p divides n.

I.10 In the case $n = 3$ Euler observes that of course

$$x^3 + y^3 = (x + y)(x^2 - xy + y^2).$$

But now he argues much as we did for $n = 2$ to conclude that x and y are both odd, allowing him to set $x = a - b$ and $y = a + b$. That yields $x^3 + y^3 = 2a(a^2 + 3b^2)$, where the factors $2a$ and $a^2 + 3b^2$ are either relatively prime or have greatest common divisor 3. Euler now has to study odd cubes of the shape $a^2 + 3b^2$. The detailed arguments were omitted from his book on algebra; later Legendre reproduced Euler's proof of Fermat's Last Theorem for $n = 3$, still with the omission of details, leading to the allegation that there was a gap in Euler's proof. It is not at all an impossible task to attempt to duplicate Euler's arguments, especially if we are willing to work with the integers $a \pm \sqrt{-3}b$, as did Gauss a little later. We will see in the next lecture that there is unique factorization in the domain of integers of the field $\mathbb{Q}(\sqrt{-3})$, and since the only units of this domain are the 6 th roots of unity, one can virtually pretend to be working with the ordinary integers $\mathbb{Z}$.

Quite a bit later, Dirichlet is forced to a similar analysis of integers of the shape $a^2 - 5b^2$. Here too, we could work comfortably with integers $a + \sqrt{5}b$, helped by the fact that the units are just the powers of $\frac{1}{2}(1 + \sqrt{5})$.

The arguments of Dirichlet and Lamé in the case $n = 7$ were soon simplified by Lebesgue, who makes clever use of the identity

$$\begin{aligned}(a + b + c)^7 - (a^7 + b^7 + c^7) &= 7(a + b)(b + c)(c + a) \\ &\quad \times((a^2 + b^2 + c^2 + ab + bc + ca)^2 + abc(a + b + c)).\end{aligned}$$

Checking such a thing is a bore, but it's fun to "discover" it. Of course we are not at all surprised that the right hand side is divisible by the prime 7 because we know that for p prime and $k \neq 1, p$ the binomial coefficients $\binom{p}{k}$ are divisible by p.

I.11 In the light of this extra detail we see that Lamé's suggestion is precisely an effort to return to working with just linear factors. The point is that, viewed as a polynomial in x, $x^p + y^p$ with p odd obviously has the zero $-y$ and hence the factor $x + y$. If $\zeta^p = 1$, then ζy is just as good a p th root of y^p as is y, so also $x + \zeta y$ is a factor. And finally, of course, if $\zeta \neq 1$ is a p th root of unity, then its powers ζ^r with $r = 0, 1, \ldots, p - 1$ give all p of the p th roots of unity.

The trouble is that the answer to Liouville's remark: "Is there not here a gap to be filled?" is a resounding "Yes, indeed". The assumption of unique factorization in the rings of cyclotomic integers is false. Still, we will see that Kummer deals with that almost completely. The word *cyclotomic* indeed pertains, because the roots of unity *cut* (as in *tom*ography) the circle.

Yet even now (in 1995), we need a conjecture of Vandiver to recover totally from this problem. However, it turns out that a much worse lacuna in Lamé's argument is his failure to appreciate that there may be nontrivial units. The matter of units in the rings $\mathbb{Z}[\zeta]$ remains one of the higher mysteries. Kummer's achievement will be to tame this matter somewhat. Of course, we too see immediately that when $\zeta^p = 1$, with p an odd prime, the $2p$ th roots of unity are units of the domain, but one has to look twice to see that $(\zeta^a - 1)/(\zeta^b - 1)$ is always a unit for $1 \le a, b \le p - 1$. Since their reciprocals are of the same shape, it is enough to check that *all* these objects are cyclotomic *integers*. But that follows from the observation that there is a positive integer k so that $a \equiv kb \pmod{p}$.

That $\zeta + \zeta^{p-1} = \zeta^{-1}(\zeta^2 + 1)$ is a unit is just a special case. It is a painful fact that in general there are yet further units not generated by the fairly accessible units described above.

I.12 I had better also make a remark about *integers*. We know that the rational numbers r/s in $\mathbb{Q}$ are exactly the roots of polynomial equations $sX - r = 0$, and that a rational number is an integer if we may take $s = 1$. In this spirit, a number is an *algebraic* number if it is a zero of a polynomial $a_0X^n + a_1X^{n-1} + \cdots + a_{n-1}X + a_n$ with rational integer coefficients (that is, with coefficients in $\mathbb{Z}$; the adjective "rational" is there for emphasis to indicate that we mean honest-to-goodness integers). And we say that an *algebraic* number is an *integer* if one may take $a_0 = 1$, that is, if its defining polynomial is monic. It is a sadness that algebraic integers may not look like integers to the naked untutored eye. For example, if the integer D is not a square, and $D \not\equiv 1 \pmod{4}$, the integers of the field $\mathbb{Q}(\sqrt{D})$ are all $\mathbb{Z}$-linear combinations of 1 and $\sqrt{D}$ — as we might expect. But if $D \equiv 1 \pmod{4}$, then $\frac{1}{2}(a + b\sqrt{D})$ is an integer whenever the integers a and b have the same parity. Already for cubic fields $\mathbb{Q}(\sqrt[3]{D})$, say, it is quite a nontrivial exercise to determine exactly when $(a + b\sqrt[3]{D} + c\sqrt[3]{D^2})/d$ is an integer. Thus it is good fortune that the integers of $\mathbb{Q}(\zeta)$ are indeed just the naïve sums $a_0 + a_1\zeta + \cdots + a_{p-2}\zeta^{p-2}$. I have cruelly omitted the term $a_{p-1}\zeta^{p-1}$ because in these remarks I am no longer too shy to admit that, of course, $1 + \zeta + \cdots + \zeta^{p-1} = 0$.

I.13 One talks about *fields* of numbers. A field — the simplest example is the field of rationals $\mathbb{Q}$ — is any system in which one may safely add and multiply normally (and subtract and divide — except for division by 0) and stay within the system. If we now append some algebraic number

α to $\mathbb{Q}$, the *extension* field $\mathbb{Q}(\alpha)$ must *inter alia* contain all nonnegative powers $1, \alpha, \alpha^2, \ldots$. It is now the essence of algebraicity that there is some smallest integer n so that no more than n of these powers are linearly independent over $\mathbb{Q}$. Then $\mathbb{Q}(\alpha)$ is, among other things, simply an n-dimensional vector space over $\mathbb{Q}$, and we say that α is algebraic of *degree* n. When p is prime the cyclotomic field $\mathbb{Q}(\zeta)$ is of degree $p-1$ over $\mathbb{Q}$.

I.14 The related remarkable fact is that if α is algebraic of degree n, then the reciprocal of a number $a_0 + a_1\alpha + \cdots + a_{n-1}\alpha^{n-1}$, with the a_i in $\mathbb{Q}$, is again of that shape. This miracle becomes more prosaic if we realize that it is just a generalization of the fact that one can clear surds from a denominator: For example, $1/(a+b\sqrt{2}) = a/(a^2-2b^2) - \sqrt{2}b/(a^2-2b^2)$.

I.15 I should add the explanation that my Dramatis Personæ is lifted direct from Hendrik Lenstra's essay, *op. cit.* with the naïve addition of the names of Frey, Ribet and Wiles. Were I starting these notes now [in 1995] I would surely include Jean-Pierre Serre and Richard Taylor, and no doubt John Tate, Barry Mazur and ...

R. P. Langlands

Photograph by C. J. Mozzochi, Princeton

LECTURE II

Der Fermatscher Satz ist zwar mehr ein Curiosum als ein Hauptpunkt der Wissenschaft.

E. E. Kummer

Lamé's argument produced a flurry of investigation in Paris on the matter of unique factorization of cyclotomic integers. To understand the underlying ideas we had better recall just why the usual integers $\mathbb{Z}$ are known to have unique factorization.

The argument goes back to Euclid and relies on observing that given integers a and b, with $b \neq 0$, there is some multiple, say kb, of b so that

$$|a - kb| < |b| .$$

Of course one picks k as the integer part of the quotient of a/b; then $a - kb$ is the *remainder*. The point is that a common divisor of a and b is also a common divisor of b and of $r = |a - kb|$. Now iterating the argument reveals the greatest common divisor $d = (a, b)$ of a and b — without one having to factorize a or b.

The process is the same thing as expanding a/b as a *continued fraction*:

$$k_0 + \cfrac{1}{k_1 + \cfrac{1}{k_2 + \cfrac{1}{\ddots + \cfrac{1}{k_{n-1} + \cfrac{1}{k_n}}}}}$$

abbreviated by writing $a/b = [k_0 , k_1 , \dots , k_{n-1} , k_n]$. Its truncations

$$[k_0 , k_1 , \dots , k_{h-1} , k_h] = p_h/q_h$$

yield rational numbers (called *convergents* because they converge rapidly to a/b, each approximating it surprisingly well), where the p_h and q_h are relatively prime, and are given by the recursive formulas

$$\begin{pmatrix} p_h & p_{h-1} \\ q_h & q_{h-1} \end{pmatrix} \begin{pmatrix} k_{h+1} & 1 \\ 1 & 0 \end{pmatrix} = \begin{pmatrix} p_{h+1} & p_h \\ q_{h+1} & q_h \end{pmatrix}$$

where

$$\begin{pmatrix} p_{-1} & p_{-2} \\ q_{-1} & q_{-2} \end{pmatrix} = \begin{pmatrix} 1 & 0 \\ 0 & 1 \end{pmatrix}.$$

It follows that $p_h q_{h-1} - p_{h-1} q_h = (-1)^{h+1}$. Since $p_n / q_n = a/b$ using the definition, we must have $a = dp_n$ and $b = dq_n$, where $d = \gcd(a, b)$ is the greatest common divisor of a and b. Hence

$$dp_n q_{n-1} - p_{n-1} dq_n = (-1)^{n+1} d = q_{n-1} a - p_{n-1} b$$

displays the gcd as a $\mathbb{Z}$-linear combination $d = sa + tb$ of a and b.

Since, in any case, the Euclidean Algorithm itself relies on the Well-Ordering Principle:

A nonempty set of positive integers contains a smallest element.

we might have appealed directly to that principle — which is well known to be equivalent to the Principle of Induction — to remark that the set

$$I = (a, b) = \{sa + tb : s, t \in \mathbb{Z}\}$$

of all $\mathbb{Z}$-linear combinations of a and b has a least positive element d, and that d is fairly readily seen to be a greatest common divisor, usually also denoted by (a, b), of a and b.

By induction it is easy to see that every positive integer is a product of positive irreducible integers (one views 1 as being the empty product). In order to see that the decompositions are unique, up to the order of the factors, suppose to the contrary that n has two essentially distinct decompositions. Of course, then n is neither irreducible nor 1, so we may set $n = ab$ with positive a and b both less than n. Our hypothesis now becomes that there is some irreducible q which divides $n = ab$ — one writes $q \mid ab$ — but neither $q \mid a$ nor $q \mid b$.

The next remarks work for the usual integers $\mathbb{Z}$, but will not necessarily work with more generalized "integers". For q in $\mathbb{Z}$ we can say that since q is irreducible, $q \nmid a$ entails that the greatest common divisor (a, q) of a and d is 1. Thus there are integers s and t, say, so that $1 = sa + tq$, whence $b = sab + tqb$, and evidently $q \mid b$. Thus every irreducible integer is *prime*, a prime being an element p which is *not* a *unit* — that is, $p \nmid 1$, p does not divide 1 — having the property that $p \mid ab$ implies $p \mid a$ or $p \mid b$. Once we know that every irreducible is prime, it follows immediately by an induction on the number of irreducible factors that decomposition into irreducibles is unique.

Now consider, for example, the Eisenstein integers $\{a + \rho b : a, b \in \mathbb{Z}\}$, where $\rho = \frac{1}{2}(-1 + \sqrt{-3})$ and so $\rho^3 = 1$. We associate with each such number its *norm*

$$\mathrm{N}(a + \rho b) = (a + \rho b)(a + \rho^2 b) = a^2 - ab + b^2 .$$

One then shows, following the idea of division with remainder, that given integers $a + \rho b$ and $c + \rho d$, there are integers $u + \rho v$ and $r + \rho s$ so that

$$a + \rho b = (u + \rho v)(c + \rho d) + (r + \rho s) \quad \text{with} \quad \mathrm{N}(r + \rho s) \leq \tfrac{3}{4}\mathrm{N}(c + \rho d).$$

Thus the Eisenstein integers form a Euclidean ring with respect to the norm and just as for $\mathbb{Z}$, it follows that they constitute a unique factorization domain. Similarly, one sees that the Gaussian integers $\{a + ib : a, b \in \mathbb{Z}\}$ have the Euclidean property with respect to their norm $N(a + ib) = a^2 + b^2$.

With rather more effort, Cauchy succeeded in guessing correctly that the domains $\mathbb{Z}[\zeta_m]$ of cyclotomic integers are Euclidean for

$$m = 5, 6, 7, 8, 9, 10, 12, 14, 15$$

as well; here $\zeta_m^m = 1$ and the ζ_m are *primitive* m th roots — no smaller power of ζ_m is 1. Mind you, this was a less comprehensive achievement than may appear. Cauchy appears to miss the fact that, when n is odd, $\zeta_n = -\zeta_{2n}$, so $\mathbb{Z}[\zeta_n] = \mathbb{Z}[\zeta_{2n}]$. In addition to these results, Cauchy also claimed to have analytic arguments suggesting that $\mathbb{Z}[\zeta_m]$ is Euclidean for all large m, say bigger than 10, but he was unable to find a decisive proof.

The explanation for this failure came from Germany, in a letter from Kummer to Liouville. (At this time German mathematics, inspired by stars like Gauss, Jacobi, and Dirichlet, was far more sophisticated than that in France.) On 28 April 1847, Kummer explained that it is not the case in general that domains $\mathbb{Z}[\zeta_m]$ have unique factorization; however he had succeeded in retrieving this property by introducing *ideal* numbers.

Suitably chastened, Cauchy showed in the next *Comptes Rendus de l'Académie des Sciences* that indeed the numbers $\mathbb{Z}[\zeta_{23}]$ do not have the property of unique factorization into irreducibles.

Once forewarned, it is easy to see that unique factorization is in fact rather unusual. Take, for example, the domain

$$\mathbb{Z}[\sqrt{-5}] = \{a + \sqrt{-5}b : a, b \in \mathbb{Z}\}.$$

We see that

$$6 = 2 \cdot 3 = (1 + \sqrt{-5})(1 - \sqrt{-5}).$$

Plainly, the irreducibles 2 and 3 are not prime; nor are the irreducibles $1 \pm \sqrt{-5}$. Kummer deals with this by introducing *ideal* numbers — the word "imaginary" is already in use — and shows that he may write, say, $2 = \mathfrak{a}\bar{\mathfrak{a}}$, $3 = \mathfrak{b}\bar{\mathfrak{b}}$, whereby the preceding decompositions become

$$6 = \mathfrak{a}\bar{\mathfrak{a}} \cdot \mathfrak{b}\bar{\mathfrak{b}} = \mathfrak{a}\bar{\mathfrak{b}} \cdot \bar{\mathfrak{a}}\mathfrak{b}.$$

The new ideal irreducibles are prime. At little more than the cost of writing a few Gothic letters, unique factorization is restored.

These ideas had not arisen in the context of Fermat's Last Theorem, but rather from a desire to generalize the *law of quadratic reciprocity* of

Gauss (1801). The law states that if p and q are distinct odd primes, then the two congruences

$$x^2 \equiv p \pmod{q} \quad \text{and} \quad y^2 \equiv q \pmod{p}$$

either both have a solution in integers x and y or are both insoluble, except when both p and q are 3 modulo 4, in which case one is solvable and the other is not. The question was to generalize this rule to powers higher than the second.

Gauss himself had shown, in 1832, that to formulate a reciprocity law for fourth powers one needed the Gaussian integers $\mathbb{Z}[i]$. He had already established a cubic reciprocity law using the numbers $\mathbb{Z}[\rho]$. His results suggested that for higher powers n, one must first deal with the following question:

> Can every prime p congruent to 1 modulo n be written as the norm of a cyclotomic integer in $\mathbb{Z}[\zeta_n]$?

Incidentally, the norm of $a(\zeta_n) = a_0 + a_1\zeta_n + \cdots + a_{n-1}\zeta_n^{n-1}$ is the product of the different $a(\zeta_n^r)$ as ζ_n^r runs through the primitive nth roots of unity. [It is not too hard to see that ζ_n^r is a primitive nth root exactly when r and n are relatively prime. The number of such r is denoted by $\phi(n)$, where ϕ is the so-called Euler totient function.]

It seems clear that Jacobi, and later Eisenstein, realized that the question just posed has an affirmative answer if and only if the domain $\mathbb{Z}[\zeta_n]$ has unique factorization. A beautiful proof of this is given by Kummer (*Collected Papers*, (New York: Springer-Verlag, 1975), Vol I, pp. 241–243). Apparently Kummer had suggested in 1844, in a paper withdrawn before publication, that the answer is yes for all n, but had made the timely discovery that this is false for $n = 23$.

The mythology of Fermat's Last Theorem claims that Kummer perpetrated Lamé's error. Battered but not defeated, he retrieved matters by his introduction of ideal numbers. Whatever, it was in 1847, just when the Parisian mathematicians were beating their heads against the wall of unique factorization, that Kummer got the idea of applying his new ideal theory to Fermat's Last Theorem.

As regards unique factorization into irreducibles, it turns out that there are just 30 different cyclotomic fields $\mathbb{Q}[\zeta_n]$ with unique factorization:

$$n = 1, 3, 4, 5, 7, 8, 9, 11, 12, 13, 15, 16, 17, 19, 20, 21, 24, 25, 27, 28,$$
$$32, 33, 35, 36, 40, 44, 45, 48, 60, 84.$$

The notion "Euclidean" plays no role in the eventual argument.* To show that these fields have unique factorization is routine; excluding all other cases is not quite. In 1979 only 13 of these were known to be Euclidean.

*J. H. Masley and H. L. Montgomery, 'Cyclotomic fields with unique factorization', *J. für Math.* **286/87** (1976), pp. 248–256.

Kummer eventually established the higher reciprocity laws. It was a task he plainly valued far more highly than his researches on Fermat's Last Theorem, to which he owes his fame.

> *Bei meinen Untersuchungen über die Theorie der complexen Zahlen und den Anwendungen derselben auf den Beweis des Fermatschen Lehrsatzes, welchen ich der Akademie der Wissenschaften vor drei Jahre mitzutheilen die Ehre gehabt habe, ist es mir gelungen die allgemeinen Reciprocitätsgesetze für beliebig hohe Potenzreste zu entdecken, welche nach dem gegenwärtigen Stande der Zahlentheorie als die Hauptaufgabe und die Spitze dieser Wissenschaft anzusehen sind.*
>
> E. E. Kummer (1850)

This lecture is substantially plagiarized from the introduction to the thesis of Hendrik Lenstra, Jr., which I had translated from the Dutch a dozen years earlier; see *Math. Intelligencer* 2 (1980), pp. 6-15, 73-77, 99-103; that work is Copyright © 1980 Springer-Verlag New York, Inc. and is partly reprinted here with permission.

Notes and Remarks

II.1 I had best say a few words about continued fractions. Everyone knows that BC — before calculators, π was 22/7 and AD — after decimals, π became = 3.14159265 . . . . In other words, π is quite well approximated — that $\pi \neq 22/7$ follows from

$$0 \neq \int_0^1 \frac{t^4(1-t)^4}{1+t^2}\,dt = \frac{22}{7} - \pi$$

— by the vulgar fraction 22/7; and some of us know that 355/113 does a yet better job since it yields as many as seven correct decimal digits. The "why this is so" of the matter is this: It happens that

$$\pi = 3 + \cfrac{1}{7 + \cfrac{1}{15 + \cfrac{1}{1 + \cfrac{1}{292 + \cfrac{1}{1 + \ddots}}}}}$$

For brevity, a commonly used flat notation for such a *continued fraction* expansion is

$$[3\,,7\,,15\,,1\,,292\,,1\,,\ldots]\,.$$

The entries 3, 7, 15, . . . are known as the *partial quotients*. Truncations, for example

$$[3\,,7] = 3 + \frac{1}{7} = \frac{22}{7} \quad \text{or} \quad [3\,,7\,,15\,,1] = \frac{355}{113}$$

are known as *convergents* of π.

The important truth is that the convergents p_h/q_h, $h = 0, 1, 2, \ldots$ yield good rational approximations, indeed excellent ones relative to the size of the denominator q_h. In the present example

$$\left|\pi - \frac{22}{7}\right| < \frac{1}{15 \cdot 7^2} \quad \text{and} \quad \left|\pi - \frac{355}{113}\right| < \frac{1}{292 \cdot 113^2}$$

instancing the general result that

$$\left|\pi - \frac{p_h}{q_h}\right| < \frac{1}{c_{h+1} \cdot q_h^2},$$

where c_{h+1} is the next (as yet unused) partial quotient. In particular $22/7$ and $355/113$ yield unusually good approximations to π because the subsequent partial quotients, respectively 15 and 292, are relatively large.

II.2 It is, of course, a serious pain to compute convergents and the like, and it is therefore a great delight for those of us who are not frightened by 2×2 matrices to show by induction, using the definition

$$[a_0, a_1, \ldots, a_n] = a_0 + 1/[a_1, \ldots, a_n],$$

that for all n

$$\begin{pmatrix} a_0 & 1 \\ 1 & 0 \end{pmatrix}\begin{pmatrix} a_1 & 1 \\ 1 & 0 \end{pmatrix}\cdots\begin{pmatrix} a_n & 1 \\ 1 & 0 \end{pmatrix} - \begin{pmatrix} p_n & p_{n-1} \\ q_n & q_{n-1} \end{pmatrix}$$

reports essentially the same information as $[a_0, a_1, \ldots, a_n] = p_n/q_n$. This correspondence reduces all continued fraction computations to just simple manipulation of the corresponding matrices.

In particular, taking determinants yields $p_n q_{n-1} - p_{n-1} q_n = (-1)^{n+1}$, which is $\dfrac{p_n}{q_n} = \dfrac{p_{n-1}}{q_{n-1}} + (-1)^{n-1}\dfrac{1}{q_{n-1}q_n}$; so that one gets

$$\frac{p_n}{q_n} = a_0 + \frac{1}{q_0 q_1} - \frac{1}{q_1 q_2} + \cdots + (-1)^{n-1}\frac{1}{q_{n-1}q_n},$$

allowing a ready proof of the convergence and approximation properties of continued fraction expansions.

Indeed, given that the partial quotients a_i are positive integers for $i \geq 1$ it follows that (q_n) is an increasing sequence of positive integers. So

$$\alpha = [a_0, a_1, a_2, \ldots]$$

exists, and moreover we have

$$\alpha - \frac{p_n}{q_n} = (-1)^n\Big(\frac{1}{q_n q_{n+1}} - \frac{1}{q_{n+1}q_{n+2}} + \cdots\Big),$$

showing the high quality of convergence yielded by the convergents; for certainly, this entails that

$$\left|\alpha - \frac{p_n}{q_n}\right| < \frac{1}{q_n q_{n+1}}.$$

II.3 The convergents p_n/q_n of α do not just provide surprisingly good approximations; they provide all the "locally best" approximations. To be precise one always has, for all integers p and q with $0 < q < q_n$, that

$$|q_n\alpha - p_n| < |q\alpha - p|.$$

The preceding remarks on continued fractions somewhat resemble remarks in my paper with Enrico Bombieri, 'Continued fractions of algebraic numbers' [in *Computational Algebra and Number Theory*, Wieb Bosma and Alf van der Poorten eds. (Dordrecht: Kluwer, 1995), 137–152], not surprisingly, as they were written at much the same time — so, naturally, I made the same jokes.

II.4 In her recent book *The Search for E T Bell*, (Washington, D.C.: Mathematical Association of America, 1993), Constance Reid highlights "Bell's Lemma" (pp. 257–258), according to which the general solution to the integer equation $xy = uv$ is given by $x = ab$, $y = cd$, $u = ac$, $v = bd$ with a, b, c, d integers and b and c relatively prime. It is by this principle that the introduction of ideal factors resolves the problem of lack of unique factorization. Incidentally, my liking for the book is materially increased by the *curiosum* whereby, after thanking various mathematicians who had provided help, Ms. Reid lists me among the members of the science fiction community who provided useful comment.

II.5 I shouldn't be too coy about the reciprocity laws. Suppose p and q are odd primes. On the surface, quadratic reciprocity is simply the fact that unless both $p \equiv -1 \pmod 4$ and $q \equiv -1 \pmod 4$, then neither or both $x^2 \equiv p \pmod q$ and $y^2 \equiv q \pmod p$ have a solution. Incidentally, one says that p is a *quadratic residue* modulo q if $x^2 \equiv p \pmod q$ — that is, if p is the remainder of a square modulo q. Otherwise, p is said to be a *quadratic nonresidue* modulo q. The higher reciprocity laws are higher degree generalizations of this rule.

Said this way, quadratic reciprocity does not make one's pulse race. The point is that it is now a fairly easy matter to solve an infinity of problems, say to find all primes q so that p is a quadratic residue modulo q, because by reciprocity this becomes the finite problem of determining appropriate classes modulo p.

There are, to put it mildly, quite a few different proofs of the law of quadratic reciprocity. The sort of thing that's going on here relies on the fact, known to Fermat, that in any case $q^{p-1} \equiv 1 \pmod p$. This entails that $q^{\frac{1}{2}(p-1)}$ must be a square root of unity, hence $\pm 1 \pmod p$. Quadratic reciprocity allows one to determine, exactly as easily as carrying out the Euclidean algorithm on the pair (q, p), just which square root applies. The value 1 or -1 is traditionally denoted by the *Legendre symbol* $\left(\frac{q}{p}\right)$. In that spirit, the higher reciprocity laws may be thought of as a matter of rules for evaluating analogous "symbols" by appropriate higher roots of unity.

In any case, Gauss's law of quadratic reciprocity was correctly seen as a jewel of number theory. There is little to wonder then at Kummer's pride in its generalization, and his declaration that this was to be viewed as a true summit of mathematics.

II.6 My reference to *J. für Math.* is meant to refer to *Journal für die reine und angewandte Mathematik*, still sometimes known as *Crelle's Journal*. Perhaps I should have written *J. für Math. (Crelle)*.

II.7 You'll have noticed my admission that I follow the start of Hendrik Lenstra's 'Euclidean number fields' fairly closely. When you actually read the original in the *Intelligencer* — and you should — you'll be scandalized by just how closely. The two rules of plagiarism are that one should admit to one's source, and that one should improve on it. But I found the latter exceptionally difficult. The story is this. Back in Amsterdam in the seventies, Lenstra had written a sheaf, indeed several sheaves, of papers but had not submitted a thesis. When ultimately cajoled into doing so, he refused to follow the standard way. Let me quote some of his words in poor translation:

> The form in which this dissertation appears represents an attempt to distinguish the two functions that a thesis traditionally fulfils — that of *scientific publication*, and that of *Festschrift on the occasion of the writer's graduation* ... .
>
> My aim in the first part of this thesis is to give the general mathematical public an impression of my subject, of the historical background against which one should view it, and of the methods employed in it. The second part consists of three articles ... . These articles are included in only a small proportion of the copies of the thesis, and then only to satisfy [the rules of the University of Amsterdam].

Naturally, in order to make it more accessible to his immediate public, Lenstra wrote the first part of his thesis in Dutch.

The fame of this wonderful essay — mind you, the tyranny of language minimized the number of possible critics — had spread far and wide by the time of the International Congress of Mathematicians, Helsinki, 1978. *The Mathematical Intelligencer* was just being launched by Springer-Verlag and it decided to commission a translation.

Don Zagier (who had taught himself Dutch as a Christmas present for his Belgian mother) was the first approached, but declined in my favor. I checked with Hendrik, who admitted that he could, of course, do the translation himself but, having put such effort into getting the words just right in Dutch, was not of a mood to try it again in English. Eventually, my translation had its grammar corrected by Hendrik, and there it was. Even more than a dozen years later I'm pleased to say I find it hard to improve on our words.

Notes on Fermat's Last Theorem

LECTURE III

Sophie Germain was a French mathematician, a contemporary of Cauchy and Legendre, with whom she corresponded. Her theorem, brought by Legendre to the attention of the illustrious members of the Institut de France, was greeted with great admiration.

P. Ribenboim

Abel's formulas. Suppose that $x^p+y^p+z^p = 0$ [and that $\gcd(x, y, z) = 1$]. Then

$$\gcd\left(x+y, \frac{x^p+y^p}{x+y}\right) = 1 \quad \text{or} \quad p$$

and if $p \mid (x^p + y^p)$, then $p \mid (x + y)$ and $p \,\Big\|\, \dfrac{x^p + y^p}{x+y}$.*

Here p is an odd prime. To see the claim, just notice that

$$\begin{aligned}\frac{x^p+y^p}{x+y} &= \frac{(x+y-y)^p+y^p}{x+y} \\ &\equiv py^{p-1} - \tfrac{1}{2}p(p-1)y^{p-2}(x+y) \pmod{(x+y)^2}\end{aligned}$$

on the one hand; while on the other hand it is $\equiv (x+y)^{p-1} \pmod{p}$. The basic underlying fact is that for any a, b,

$$(a+b)^p = a^p + b^p \pmod{p};$$

indeed, this characterizes primes p. But more than that, there is Fermat's Theorem (often referred to as his *little* theorem) according to which one has $a^p \equiv a \pmod{p}$ for any rational integer a. So for a collection of integers $a_1, a_2, \ldots, a_n$ we have the congenial result that

$$(a_1 + a_2 + \cdots + a_n)^p \equiv a_1 + a_2 + \cdots + a_n \pmod{p}.$$

---

* p "exactly" divides ... ; that is, p divides the quantity but p^2 does not.

In any case, by the factorizations $-z^p = (x+y)\dfrac{x^p+y^p}{x+y}, \ldots$, we have

$$\begin{array}{lll} x+y=c^p & \dfrac{x^p+y^p}{x+y}=\gamma^p & \text{if}\quad p\nmid z, \\ y+z=a^p & \dfrac{y^p+z^p}{y+z}=\alpha^p & \text{if}\quad p\nmid x, \\ z+x=b^p & \dfrac{z^p+x^p}{z+x}=\beta^p & \text{if}\quad p\nmid y. \end{array}$$

So in the *first case* of Fermat's Last Theorem, when $p \nmid xyz$ we have, say $2x = -a^p + b^p + c^p$.

Incidentally, if $p \mid z$ then $x + y = p^{p-1}c^p$ and $\dfrac{x^p+y^p}{x+y} = p\gamma^p$ with $p \nmid \gamma$.

Now Sophie Germain argues as follows: If $x^p + y^p + z^p = 0$ then

$$x^p + y^p + z^p \equiv 0 \pmod{q}$$

for every q. Suppose now that $q = 2p + 1$ happens to be prime. Then $q \nmid x$, say, entails $x^p \equiv \pm 1 \pmod{q}$; and similarly for y and z. But all of x^p, y^p, $z^p \equiv \pm 1$ is impossible. So, necessarily $q \mid xyz$. Say $q \mid x$. Then $-a^p + b^p + c^p \equiv 0 \pmod{q}$, so $q \mid abc$; in fact, one sees that $q \mid a$. Thus $y \equiv -z$, so $\alpha^p \equiv p y^{p-1} \pmod{q}$. Also, of course, $y^p \equiv y^{p-1} \pmod{q}$. Thus p is a p th power modulo q, which entails $p \equiv \pm 1 \pmod{q}$. But this contradicts $2p + 1 = q$. Using these ideas Legendre generalized the result to $q = 2kp + 1$ prime for several further values of k, enabling all cases $p < 100$ to be dealt with. In 1823 this was a remarkable advance.

No one doubts that there are infinitely many primes p for which also $2kp + 1$ is prime for some suitable small k, but such results remain as yet inaccessible. Nonetheless, fairly recently Adleman and Heath-Brown *inter alia* used work of Fouvry and a generalization of Sophie Germain's result to prove that the first case of the FLT holds for infinitely many prime exponents.

Sophie Germain employed her own version of the formulas I mention. Suppose that $0 < x < y < z$. Abel considered the case of Fermat's equation $x^n + y^n = z^n$ when one of x, y, or z is a prime power. Abel showed that x must be the prime power, and that if x is prime, then $z = y + 1$, that the exponent n must be a prime p, and that $p \mid y(y+1)$.

However his proof that the equation

$$x^p + y^p = (y+1)^p$$

has no nontrivial solutions for $p > 2$ was shown to be faulty some seventy years later, by Markoff. Thus even that this very special case of Fermat's equation — Abel's equation — has no solutions remained unproved until several weeks ago. For all we knew, $z - y = 1$ was possible in the FLT. On the other hand, some 20 years ago, Kustaa Inkeri and I used Baker's method to prove that $y - x$ must be large relative to the exponent if

Fermat's equation was to have a solution. Related methods had been used in preceding years by Inkeri *et al.* to show that in any putative solution to Fermat's equation, each of x, y, and z would have to be very large indeed.

Kummer's results dealt with both the first and second cases but were incongenial for extended computation. To the contrary, the results of Mirimanoff and Wieferich at the turn of the century, and a little later, of Fürtwangler, seemed almost to dispose of the first case. They showed that if there is a first case solution for exponent p, then

$$p^2 \mid (2^{p-1} - 1) \quad \text{and} \quad p^2 \mid (3^{p-1} - 1).$$

These results relied on a complicated analysis of the conditions that had first been proved by Kummer. At first no instances of p satisfying the first condition were known. Then in 1913 Meissner noticed the example $2^{1092} \equiv 1 \pmod{1093^2}$ and in 1922 Beeger found that $3511^2 \mid (2^{3510} - 1)$. Computer checks of some 20 years ago showed that there are no other cases for the base 2 with exponent less than 3×10^9. For the base 3 one has $11^2 \mid (3^{10} - 1)$; the next case is $3^{1006002} - 1 \equiv 0 \pmod{1006003^2}$.

Further such criteria for additional bases were derived subsequently. It is surely most improbable that even just the two criteria mentioned are ever satisfied simultaneously. Whatever, we have no understanding of the *Fermat quotients* $q_p(a) = (a^p - a)/p$ at all. Their divisibility by p appears to us as no more than a statistical accident.*

Appendices

Fermat's Theorem. Since the object of these notes is mainly to chat about Fermat's Last Theorem, the least I can do is to say a few words about a result that properly bears his name. Throughout p is a rational prime.

The claim $n^p \equiv n \pmod{p}$ is obvious by induction since one has that $(n+1)^p \equiv n^p + 1 \pmod{p}$.

A rather more to the point proof is to notice that $1, 2, \ldots, p-1$ are all the different remainders $\pmod{p}$ that are relatively prime to p. But if a is not zero $\pmod{p}$ then the numbers $a, 2a, \ldots, (p-1)a$ are different and nonzero mod p. Hence modulo p they must be just a permutation of $1, 2, \ldots, p-1$. Thus modulo p we have that $(p-1)! \equiv (p-1)!a^{p-1}$ and Fermat's Theorem is evident.

---

*Apropos Meissner, Landau's *Zahlentheorie* points out that with $p = 1093$ and all congruences $\pmod{1093^2}$, clearly $3^7 = 2187 = 2p + 1$ so that $3^{14} \equiv 4p + 1$. Just so, $2^{14} = 16384 = 15p - 11$, whence $2^{28} \equiv -330p + 121$. Thus $3^2 \cdot 2^{28} \equiv -2970p + 1089 \equiv -2969p - 4 \equiv 310p - 4$ $3^2 \cdot 2^{28} \cdot 7 \equiv 2170p - 28 \equiv -16p - 28$. So $3^2 \cdot 2^{26} \cdot 7 \equiv -4p - 7$ and $3^{14} \cdot 2^{182} \cdot 7^7 \equiv -(4p+7)^7 \equiv -7 \cdot 4p \cdot 7^6 - 7^7$. Thus $3^{14} \cdot 2^{182}$ is just $-4p - 1 \equiv -3^{14}$. Hence $2^{182} \equiv -1$ and Meissner's observation follows. But is $3^{1092} \equiv 1$ as well? Clearly no, because for any p both $x^{p-1} \equiv y^{p-1} \equiv 1 \pmod{p^2}$ as well as $x + y = mp$ with $p \nmid m$ are impossible. Now take $x = 3^7$, $y = -1$ to see that $3^{1092} \not\equiv 1 \pmod{1093^2}$.

It follows that there is an a' in the set of remainders $\{1, 2, \ldots, p-1\}$ so that $aa' \equiv 1 \pmod{p}$. Of course one writes $a' = a^{-1}$ and notes that $a^{-1} \equiv a^{p-2} \pmod{p}$.

It is also fun to remark that the polynomial $X^p - X$ has distinct zeros modulo p since it and its derivative $pX^{p-1} - 1 \equiv -1 \pmod{p}$ are trivially relatively prime. Hence, since a polynomial modulo a prime still has no more zeros than its degree, it follows that

$$X^p - X \equiv X(X-1)(X-2)\cdots(X-(p-1)) \pmod{p}$$

and *inter alia* — among other things — we have Wilson's Theorem

$$(p-1)! \equiv -1 \pmod{p}.$$

It is a fairly straightforward corollary to Fermat's Theorem that if a prime p divides the nth Fermat number $F_n = 2^{2^n} + 1$ then $p = k \cdot 2^{n+1} + 1$ for some integer k. Thus, as did Euler, but unlike Fermat — who must have made an arithmetical blunder — we can readily find a divisor of F_5.

Bernoulli numbers. Even engineers know about the power series

$$\sin z = z - \tfrac{1}{3!}z^3 + \tfrac{1}{5!}z^5 - \cdots \quad \text{and} \quad \cos z = 1 - \tfrac{1}{2!}z^2 + \tfrac{1}{4!}z^4 - \cdots$$

but what is the series expansion for $\tan z$?

The trick is first to notice that $f(z) = z/(e^z - 1) + \frac{1}{2}z$ is an even function, that is, $f(-z) = f(z)$, and then to write — noting that $B_3 = B_5 = B_7 = \cdots = 0$,

$$\frac{z}{e^z - 1} = \sum_{k=0}^{\infty} \frac{1}{k!} B_k z^k = 1 - \tfrac{1}{2}z + \tfrac{1}{2!}B_2 z^2 + \tfrac{1}{4!}B_4 z^4 + \tfrac{1}{6!}B_6 z^6 + \cdots.$$

After a tiny amount of work it is then clear that the Bernoulli numbers B_k can be computed by the inductive definition†

$$\binom{k+1}{1}B_k + \binom{k+1}{2}B_{k-1} + \cdots + \binom{k+1}{k}B_1 + B_0 = 0$$

with $B_0 = 1$. So $B_1 = -\frac{1}{2}$, $B_2 = \frac{1}{6}$, $B_4 = -\frac{1}{30}$, ... Obviously, the B_k all are rational numbers. However their numerators soon grow at a furious rate: indeed, because the cited series evidently has radius of convergence 2π, we must have $|B_{2k}| \sim (2k)!/(2\pi)^{2k}$.

What this is all about is the calculus of finite differences, now unsung, but still taught in applied maths courses when I was littler. Here the difference operator $\Delta : f(x) \mapsto f(x+1) - f(x)$ and its inverse Δ^{-1}, the indefinite sum, occur in place of the derivative $D : f(x) \mapsto f'(x)$ and its inverse D^{-1}, the indefinite integral. In this context one defines the Bernoulli polynomials $B_n(x)$ by

$$D(\Delta^{-1}x^n) = B_n(x), \quad n = 0, 1, 2, \ldots.$$

†Symbolically, this can be elegantly summed up as $(B+1)^{n+1} - B^{n+1} = 0$. Of course, at the end of it all one interprets B^k as B_k.

But on expanding $f(x+1)$ as a Taylor series about x, we see that

$$\Delta f(x) = \sum_{n=1}^{\infty} \frac{f^{(n)}(x)}{n!} = \sum_{n=1}^{\infty} \frac{D^n}{n!} f(x) = (e^D - 1) f(x)$$

so

$$D\Delta^{-1} = \frac{D}{e^D - 1}$$

and

$$D\Delta^{-1} x^n = \sum_{m=0}^{\infty} \frac{B_m}{m!} D^m x^n = \sum_{m=0}^{n} \binom{n}{m} B_m x^{n-m}$$

and in particular $B_n = B_n(0)$.

Of course, $\Delta^{-1} x^n$ "means" $1^n + 2^n + \cdots + (x-1)^n$ because

$$\Delta\Big(\sum_{m=1}^{x-1} m^n\Big) = x^n$$

whence

$$\Delta^{-1} x^n = \sum_{m=1}^{x-1} m^n = \frac{1}{n+1}\big(B_{n+1}(x) - B_{n+1}(1)\big).$$

We can now readily see the formal version of the useful Euler-Maclaurin formula

$$f(x) = \int_x^{x+1} f(t)\,dt + \sum_{n=1}^{\infty} \frac{B_n}{n!} \Delta D^{n-1} f(x)$$

allowing one to compare sums with integrals.

I'm going to religiously confine each of the Lectures to just four pages* so there's just no room for $\tan z$; but surely $iz \coth iz = z \cot z$, and then $\tan z = \cot z - 2\cot 2z$, gives enough of a hint. Worse though, there is then no space to discuss Euler's evaluation of the special values of the Riemann ζ-function; that is, of the sums

$$\zeta(2k) = \sum_{n=1}^{\infty} \frac{1}{n^{2k}} = 1 + \frac{1}{2^{2k}} + \frac{1}{3^{2k}} + \frac{1}{4^{2k}} + \cdots = (-1)^{k-1} \frac{(2\pi)^{2k}}{2(2k)!} B_{2k}.$$

Most of this lecture is just things I knew, but in the late seventies I was helped by access to some of the draft lectures that became Paulo Ribenboim's *13 Lectures on Fermat's Last Theorem*, (New York: Springer-Verlag, 1979).

*Well, that remark from the original handout doesn't make a great deal of sense now, having fallen foe to different fonts and resizing of the pages, but I'll stick with its implications regardless. I postpone Euler's evaluation of $\zeta(2k)$ to a footnote in Lecture VII.

Notes and Remarks

III.1 The largest *known* "Sophie Germain prime" p (such that $2p+1$ also is prime) is $2687145 \cdot 3003 \cdot 10^{5072} - 1$. It was found recently by Harvey Dubner, in October 1995.

III.2 Pseudoprimes. It is not the case that N is prime if $2^N \equiv 2 \pmod N$, as is illustrated by the composite number $N = 341 = 11 \times 31$. Such N, using Fermat's theorem to pretend to be primes, are called *pseudoprimes* (to the base 2). Nonetheless, Fermat's theorem does provide a primitive primality test in that any N which does not succeed even in pretending to be a pseudoprime, is assuredly composite. One might fear that the hefty size of numbers 2^N reduces such a test purely to theory. But in fact, we just need quantities modulo N, and then it's not hard to see that we need never deal with numbers greater than N^2. Indeed, writing $N = b_n b_{n-1} \ldots b_1 b_0$ in binary notation — so the b_i are 0 or 1 — we can compute $a^N \pmod N$ sequentially as

$$\Big(\cdots((a^2 \times a^{b_{n-1}})^2 \times a^{b_{n-2}})^2 \times \;\cdots\; \times a^{b_1}\Big)^2 \times a^{b_0}.$$

If we reduce modulo N after each operation, then we never have to do anything more drastic than to square mod N or to multiply by a mod N.

III.3 Carmichael numbers. Mind you, there are composite numbers N that are pseudoprimes with respect to every base. That is,

$$x^N \equiv x \pmod N$$

for every integer x. A number N is Carmichael if it is squarefree and if for every prime p dividing N we have that $(p-1) \mid (N-1)$. The smallest example is 561. It was recently shown by W. R. Alford, Andrew Granville, and Carl Pomerance that 'There are infinitely many Carmichael numbers' [*Ann. Math.* **140** (1994), pp. 703–722]. Andrew Granville provides a nice summary on 'Primality testing and Carmichael numbers' in *Notices Amer. Math. Soc.* [**39** (1992), pp. 696–700]. A recent issue of *Nieuw Archief voor Wiskunde* [4e Serie, **11** (1993), pp. 199–209] carries a delightful introductory article 'Carmichael numbers', by Carl Pomerance.

III.4 The multiplicative version of the proof outlined for Fermat's theorem immediately yields Euler's generalization whereby

$$a^{\phi(N)} \equiv 1 \pmod N$$

for a relatively prime to N. Here the Euler totient function $\phi(N)$ is the number of positive integers less than N and prime to it. The point is, just as the $\phi(p) = p - 1$ nonzero remainders modulo p form a group under multiplication, so the $\phi(N)$ remainders prime to N form a multiplicative group modulo N. It's interesting to verify that

$$\sum_{d \mid N} \phi(d) = N.$$

We'll see later that this useful fact readily leads to the evaluation

$$\phi(N) = N \prod_{p|N} \left(1 - \frac{1}{p}\right).$$

III.5 Concerning Wilson's Theorem that $(p-1)! \equiv -1 \pmod{p}$ if and only if* p is prime, Gauss (*Disquisitiones Arithmeticæ*, Art. 76) makes the following instructive though sour remark:

> *Theorema hoc elegans... primum a cel. Waring est prolatum armigeroque Wilson adscriptum, [*Meditt. algebr. Ed. 3, p. 380*]. Sed neuter demonstrare potuit, et cel. Waring fatetur demonstrationem eo difficiliorum videri, quod nulla* notatio *fingi possit, quae numerorum primum exprimat. — At nostro quidem iudicio huiusmodi veritates ex* notionibus *potius quam ex hauriri debebant.*

That is: This elegant theorem ... is first formulated by the celebrated Waring and is ascribed by him to his assistant [his "armour-bearer"] Wilson. But neither could prove it and the celebrated Waring admits that the proof seems too difficult because there is no notation for denoting an arbitrary prime. My view is, however, that truths of this sort flow from *notions* rather than from *notations*.

III.6 It's not hard to convince oneself that a polynomial defined over a field cannot have more distinct zeros than its degree, and it's interesting to deduce from this the useful truth that every finite multiplicative subgroup of a field is cyclic. By the way, the contrast is with an equation such as $X^2 \equiv 1 \pmod{8}$, which has four distinct zeros mod 8. From this we learn that $\mathbb{Z}/8\mathbb{Z}$ is not a field, whence 8 cannot be a prime. The useful truth tells us that there always is a generator g of all the nonzero elements modulo p; that is, each such element is a power of g, and $g^n \equiv 1 \pmod{p}$ only if $p-1 \mid n$. Such a g is called a *primitive root* modulo p.

It follows that if $p \equiv 1 \pmod{4}$, then $x \equiv g^{(p-1)/4} \pmod{p}$ is of exact order 4, so $x^2 \equiv -1 \pmod{p}$. So for primes of the shape $p = 4m+1$, we see that in the ring $\mathbb{Z}[i]$ of Gaussian integers, $p \mid (x^2+1) = (x+i)(x-i)$, while plainly, p divides neither factor. So p is *not* a prime in $\mathbb{Z}[i]$, and since the Gaussian integers do have unique factorization into primes, we must have $p = (a+ib)(a-ib) = a^2+b^2$, proving that each prime of the shape $p = 4m+1$ is indeed the sum of two squares.

III.7 Classically, however, one simply noted that, at any rate, $p \mid (x^2+1)$ says that some positive integer multiple of p is the sum of two squares, say $kp = x^2+1$, and $k < p$. One then proceeds to show, by descent, that

---

*The phrase "if and only if" is almost universally abbreviated as "iff" in casual mathematical use. Credit for introducing this felicitous notation is claimed by Paul Halmos (see his autobiography). On the other hand, journals invariably enjoin us to "eschew abbreviations such as 'iff' ... "; apparently we must then refer to the Canadian Rockies resort as "Banf and only Banf".

therefore p is itself the sum of two squares. Indeed, suppose that $m > 1$ is the minimal positive integer so that

$$mp = a^2 + b^2$$

is a sum of two squares. Of course, $m \le k < p$. Now use the Euclidean algorithm to find remainders a' and b', both less than $\frac{1}{2}m$, by $a = mq + a'$ and $b = mr + b'$. Then

$$mp = m^2(q^2 + r^2) + 2m(qa' + rb') + a'^2 + b'^2$$

and we see that some multiple $m'm$ of m, with $m' < m$, is necessarily a sum $m'm = a'^2 + b'^2$ of two squares. Hence

$$m'm^2p = (a'^2 + b'^2)(a^2 + b^2) = (a'a + b'b)^2 + (a'b - ab')^2 .$$

But both $c = (a'a + b'b)/m$ and $d = (a'b - ab')/m$ are integers, so we have $m'p = c^2 + d^2$, thereby completing the proof using descent.

III.8 The core of this argument is the fact that the product

$$(a'^2 + b'^2)(a^2 + b^2)$$

is itself, after substituting $A = a'a + b'b$ and $B = a'b - ab'$, the quadratic form $A^2 + B^2$. This is the simplest instance of composition of quadratic forms, central to the *Disquisitiones Arithmeticæ* of Carl Friedrich Gauss. It is also an elementary manifestation of the ideas delightfully discussed by David Cox in his *Primes of the Form $x^2 + ny^2$: Fermat, Class Field Theory, and Complex Multiplication* (New York: Wiley–Interscience, 1989).

III.9 Fermat numbers. Clearly a number $2^m + 1$ cannot be prime if m has an odd factor t, for then it has an algebraic factorization*: it is easy by simple algebra to see that $2^{m/t} + 1$ is a factor. Thus the only numbers of this shape that can possibly be prime are the Fermat numbers $F_n = 2^{2^n} + 1$. Indeed $F_0 = 3$, $F_1 = 5$, $F_2 = 17$, $F_3 = 257$ and $F_4 = 65537$ all are prime. Fermat repeatedly expressed the view that all the Fermat numbers are prime, but clearly his belief rested on an arithmetical blunder. For it could have been obvious to him that $641 \mid F_5$.

Let me say a little more about the Fermat numbers $F_n = 2^{2^n} + 1$ and their factors. Plainly, we have $2^{2^{n+1}} \equiv 1 \pmod{p}$, for any prime p dividing F_n. On the other hand, by Fermat's theorem we know that the *order* of 2 modulo p is a divisor of $p - 1$. So certainly $2^{n+1} \mid (p - 1)$, which is to say that there is some integer k so that $p = k \cdot 2^{n+1} + 1$. In fact, k must be even. A little more ingeniously one notices that in these circumstances the square root of 2 (mod p) is $2^{2^{n-2}}(2^{2^{n-1}} - 1) \pmod{p}$ — a good exercise

*In the trade, slightly more sophisticated quasi-mechanical factorizations are sometimes called *Aurifeuillian*. It sounds very educated to refer to them in this way. However, to be allowed to use the word one must first pass the test of completely factorizing $2^{58} + 1$; 5 is the trivial factor and there are two more prime factors. Remarkably, I found this challenge in an "ancient" tutorial book of high school arithmetic problems: *The Tutorial Arithmetic* by W. P. Workman and R. H. Chope (London: University Tutorial Press Ltd, 1934), p. 490, Problem **53** (3). This was admittedly one of the "Harder Problems".

for high school students learning to handle exponents — and of course its order is $2^{n+2} \pmod{p}$. Hence we must have $2^{n+2} \mid (p-1)$ and we learn that a prime divisor p of F_n is necessarily of the shape $p = k \cdot 2^{n+2} + 1$. Since $257 = F_3$ cannot divide F_5 because different Fermat numbers are obviously relatively prime, the next prime, the successful divisor 641, is actually the first serious candidate.

No further prime Fermat numbers have been found and several have been completely factorized — a highly nontrivial task since the Fermat numbers grow at a frantic clip. On the evidence, and on heuristic considerations, it is not at all unreasonable to expect that no further Fermat numbers are prime.

The paper A. K. Lenstra, H. W. Lenstra Jr., M. S. Manasse, and J. M. Pollard, 'The factorization of the ninth Fermat number' [*Math. Comp.*, **61** (1993), pp. 319–349] provides a fine introduction to the Fermat numbers, a fascinating discussion of the methods used to factorize them, and copious references.

III.10 The Fermat numbers do arise in another interesting context. It was an old problem, essentially completely solved by Gauss, to determine all regular polygons that can be constructed using straight-edge and compass alone. For example, every child knows that having drawn a circle, one can use one's compass with radius unchanged to mark all the six vertices of a regular hexagon. It turns out that the only possibilities for constructible regular n-gons are with $n = 2^m s$, where s is a product of distinct Fermat primes. The point is that these are the only numbers n having the property that $\phi(n)$ is a power of 2. This is relevant because, as is not too difficult to see, points that can be reached by the classical construction method must have their coordinates in a number field obtainable from $\mathbb{Q}$ by a sequence of quadratic extensions. Extensive details of the underlying elementary algebra are given in *Abstract Algebra and Famous Impossibilities* by Arthur Jones, Sydney A. Morris and Kenneth R. Pearson, (New York: Springer-Verlag 1991).

III.11 My remarking about the "famous impossibilities" reminds me of an infamous note in *Parade Magazine* (Marilyn vos Savant, 'Ask Marilyn', 21 November 1993, p. 16) which caused some mild mathematical furore at the time. Marilyn chooses to suggest that if one may not trisect an angle with markings on one's rule [not "rule*r*" — I remember my teachers shouting that the Queen is our *ruler*, but straight lines are drawn using one's *rule*], then hyperbolic geometry is out in proving the FLT. To this I say "Up a gum tree" [that being one of those teachers' response to our speaking irrelevant nonsense]. One may prove the FLT any old way that one cares, provided the mathematics is correct.

There is the subsidiary question of doing it the way Fermat did. But that, no doubt, has been done by thousands. One makes a foolish mistake and realizes it the next morning. But one forgets to rub out one's claim in the margin, and one forgets to tell one's son not to reprint one's marginal scribbles uncritically.

I am indebted to Rodolfo Ruiz-Huidobro (Rudy Ruiz) for sending me the offending *Parade* article; I had seen it discussed, with somewhat exaggerated passion, in the e-math Fermat discussion group a little earlier.

III.12 By the way. John Coates reports (conversation with me in Hong Kong, December, 1993) having actually seen, in the library of Emmanuel College, a copy of Bachet's *Diophantus* of 1621 and having been quite startled by the extraordinary width of its margins.

III.13 It took three authors, Everett Howe, Hendrik Lenstra, and David Moulton, to perpetrate the marginally relevant remark:

> *"My butter, garçon, is writ large in!"*
> *a diner was heard to be chargin'.*
> *"I* had *to write there,"*
> *exclaimed waiter Pierre,*
> *"I couldn't find room in the margarine."*

III.14 An irritating aspect of the eccentric remarks of the *Parade* article is that they might well encourage the totally berserk. Should you ever need to dismiss the complete nonsense of such people, here is a wonderful retort quoted by Underwood Dudley in *Mathematical Cranks* (Washington, D.C.: Mathematical Association of America ,1992):

> The twitterings of sparrows and the writings of great scientists have one thing in common. Both are incomprehensible to the average person. One should not conclude, however, that the sparrows are too profound for ordinary human beings to understand their thoughts. They just don't have any thoughts in the ordinary human sense of the term. I regret to inform you that these few sentences of yours are precisely that type.

III.15 Mersenne primes. Primes of the shape $2^m - 1$ must have m prime. These numbers were the object of avid study and speculation by Fermat's contemporary, Marin Mersenne. They reacquired notoriety because of a relatively efficient test for the primality of numbers of shape $2^m - 1$. In consequence, it is now not uncommon to test the integrity of supercomputers by allowing them to try to find what will be the largest known prime number, and to confirm earlier results. At the time of writing, the known Mersenne primes are $M_p = 2^p - 1$ with

$$p = 2, 3, 5, 7, 13, 17, 19, 31, 61, 89, 107, 127,$$
$$521, 607, 1279, 22032281, 3217, 4253, 4423, 9689, 9941, 11213, 19937,$$
$$21701, 23209, 44497, 86243, 110503, 132049, 216091, 756839.$$

Incidentally, the first line consists of those cases found by hand; the rest were found with computer aid. While human beings are credited with all the previous cases, at least for programming the computer, in the last* case $p = 756839$ my notes reveal only that the discovery was made by a CRAY-2 in England.

Mersenne's allegation that $2^{67} - 1$ is prime was disproved† by Cole, who displayed $2^{67} - 1 = 193707721 \times 761838257287$ after working "three years of Sundays" to discover the fact. The story goes that Cole's lecture reporting the factorization consisted of his evaluating $2^{67} - 1$, and then laboriously multiplying $193707721 \times 761838257287$; *et voilà!*

III.16 The traditional fascination with Mersenne primes arises from a result going back to Euclid, namely that precisely the numbers $2^{p-1}M_p$ are both even and *perfect*: each is the sum of its *aliquot parts*, that is, of its divisors other than itself. No odd perfect numbers are known; if one exists (as it well might) it is fairly large, at any rate greater than 10^{150}.

To discuss Euclid's result (I should credit Euler with the argument) I need the function $\sigma(n) = \sum_{d|n} d$, the sum of the divisors of n. Then n is perfect exactly if $\sigma(n) = 2n$, since $\sigma(n)$ also counts n itself. Let me take it as known or obvious that σ is actually *multiplicative*, that is, that $\sigma(m)\sigma(n) = \sigma(mn)$ provided that m and n are relatively prime. Now suppose that $n = 2^{a-1}m$ is an even perfect number; with m odd. Plainly, $\sigma(2^a) = 1 + 2 + 2^2 + \cdots + 2^{a-1} = 2^a - 1$, so $\sigma(n) = (2^a - 1)\sigma(m) = 2n = 2^a m$. It's not a big thing to conclude that the only possibility is to have that $\sigma(m) = 2^a$ and $m = 2^a - 1$. But this entails that $2^a - 1$ must be prime — plainly only a prime q has the property that $\sigma(q) = q + 1$. It follows that $2^a - 1$ is a Mersenne prime, as claimed.

III.17 We're right on the edge here of what is known as "recreational mathematics". Are there, for instance, pairs m and n of integers so that $\sigma(n) = 2m$ and $\sigma(m) = 2n$? Indeed there are; such pairs are called *amicable*. Coaxing one's computer to produce large examples is good clean fun and somehow seems more profound or absolute than doing the weekend crossword. In any case, this particular question, exemplified by the pair 284 and 220, allegedly goes back to Pythagoras. Its age warrants respect, such as that we give our seniors.

III.18 Recall that $\zeta(s) = \sum n^{-s}$. There is a delightful elementary proof that $\zeta(2) = \pi^2/6$, attributed to Calabi. If we expand $(1 - x^2y^2)^{-1}$ as a geometric series and integrate termwise, then we see that

$$\int_0^1\int_0^1 (1 - x^2y^2)^{-1}\,dx\,dy = 1^{-2} + 3^{-2} + 5^{-2} + \cdots = (1 - 2^{-2})\zeta(2).$$

The clever substitution $(x, y) = (\sin u/\cos v, \sin v/\cos u)$ has Jacobian $1 - x^2y^2$ and thus maps the open triangle $T = \{u, v > 0; u + v < \pi/2\}$

---

*The next Mersenne prime, with $p = 859433$, was announced in January 1994.

†Well, it had been known for some time that it's composite, but the present sentence reads more nicely.

bijectively to the interior of the square $S = [0,1] \times [0,1]$. Hence

$$\iint_S (1 - x^2y^2)^{-1}\, dx\, dy = \iint_T du\, dv = \frac{\pi^2}{8}.$$

Calabi's idea can be generalized to give the sums of both the series

$$\sum_{n=0}^{\infty} \frac{1}{(2n+1)^{2k}} = (1 - 2^{-2k})\zeta(2k) \quad \text{and} \quad \sum_{n=0}^{\infty} \frac{(-1)^k}{(2n+1)^{2k+1}}.$$

This is elegantly detailed in Frits Beukers, Eugenio Calabi and Jonathan A. C. Kolk, 'Sums of generalized harmonic series and volumes' [*Nieuw Arch. Wisk.* 4e ser. **11** (1993), pp. 216–224].

Barry Mazur at the Boston meeting, August 1995

Photograph by C. J. Mozzochi, Princeton

LECTURE IV

Again in 1850 the Académie des Sciences de Paris offered a golden medal and a prize of 3000 *Francs to the mathematician who would solve Fermat's problem.*

In 1856 it determined to withdraw the question from competition but to award the medal to Kummer "for his beautiful researches on the complex numbers composed of roots of unity and integers".

P. Ribenboim, in part translating Cauchy.

All we really will need about ideal numbers is that, according to their purpose, there is unique factorization into ideals in every number field. But there is a bonus. An element α corresponds to the *principal* ideal (α), and so do its *associates* $\eta\alpha$, where η is any unit. Thus ideals hide units, as it were, which is great. Of course, *vice versa*, the ideal (α) corresponds to one or other of the *associates* $\eta\alpha$ of α, where η is some unit; that's not so good. But Kummer copes.

An odd prime p is *regular prime* if, roughly speaking, one can do common sense arithmetic — as regards Fermat's Last Theorem — in the field $\mathbb{Q}(\zeta)$, where $\zeta = \zeta_p$ is a primitive p th root of unity. In his first arguments of 1847 Kummer works subject to the assumptions that the p th power of an ideal can be treated as a number in the field and the truth of a lemma concerning p th powers of units that I mention below. A little later he found that both of these conditions follow from the regularity of p. We recall that in the p th cyclotomic field the integers are

$$a_0 + a_1\zeta + \cdots + a_{p-2}\zeta^{p-2}$$

with the $a_i \in \mathbb{Z}$. Our object is to demonstrate the impossibility of

$$x^p + y^p + z^p = 0$$

in nonzero rational integers x, y and z.

Case I. p regular (Kummer). We begin with the first case of the FLT, thus when $p \nmid xyz$. Then

$$x^p + y^p = \prod_{j=0}^{p-1}(x + \zeta^j y) = -z^p$$

and the factors $x + \zeta^j y$ are relatively prime, since their only possible common divisor is a factor of p, while $p \nmid z$. Hence each is a p th power — technically, each *principal ideal* $(x + \zeta^j y)$ is the p th power of an ideal of $\mathbb{Q}(\zeta)$. When p is regular it follows that we have

$$x + \zeta y = \eta \alpha^p ,$$

where α is an integer and η a unit of $\mathbb{Q}(\zeta)$; compare $36 = -4 \times -9$ and note that -4 is a square times a unit -1, not just a square.

The difficulty is that $\mathbb{Q}(\zeta)$ has nontrivial units η. Nevertheless, Kummer succeeded in proving that $\eta/\overline{\eta}$ is always a p th root of unity, ζ^{2a} say.

Now if

$$\alpha = a_0 + a_1\zeta + \cdots + a_{p-2}\zeta^{p-2} \quad \text{then} \quad \alpha^p \equiv a_0 + a_1 + \cdots + a_{p-2} \pmod{p}$$

so that $\alpha^p \equiv \overline{\alpha}^p \pmod{p}$, and we have

$$x + \zeta y \equiv (x + \overline{\zeta} y)\zeta^{2a} \quad \text{or} \quad x(1 - \zeta^{2a}) + y(\zeta - \zeta^{2a-1}) \equiv 0 \pmod{p} .$$

But the latter is impossible if $p \nmid xy$, unless $2a \equiv 1 \pmod{p}$, in which case $p \mid (x - y)$. But then, by symmetry, we can also establish $p \mid (y - z)$ and $p \mid (z - x)$; hence

$$0 = x^p + y^p + z^p \equiv 3x^p \equiv 3x \pmod{p} ,$$

whence $p \mid x$ if $p > 3$. That is a contradiction.

Finally, for completeness, FLT I with $p = 3$ is impossible by Sophie Germain. One need only note that all cubes not divisible by $7 = 2 \cdot 3 + 1$ are $\equiv \pm 1 \pmod{7}$.

Case II. p regular (Kummer). Suppose without loss of generality that $p \mid z$, and put $\lambda = 1 - \zeta$. Recall that the significance of λ is that the principal ideal (λ) satisfies $(\lambda)^{p-1} = (p)$. Suppose also that we already have the Case I result for integers α, β, γ of $\mathbb{Q}(\zeta)$; thus we are supposing that there is no nontrivial solution of

$$\alpha^p + \beta^p + \gamma^p = 0 \quad \text{if} \quad (p, \alpha\beta\gamma) = 1 .$$

We consider the apparently more general equation

$$\alpha^p + \beta^p = \varepsilon \lambda^{np} \gamma^p$$

with ε a unit and α, β, γ nonzero integers of $\mathbb{Q}(\zeta)$. We will use descent on n, and so may suppose that α and β are prime to λ.

First, one remarks that we may further suppose that α and β each have the property of being congruent to a rational integer modulo λ^2. The idea is that since we are only given α^p, there is no loss in replacing α by $\zeta^k \alpha$.

That there is no loss of generality follows because if $\zeta = 1 - \lambda$, we have $\zeta^k \equiv 1 - k\lambda \pmod{\lambda^2}$. So if $\alpha \equiv l + m\lambda \pmod{\lambda^2}$, then

$$\zeta^k \alpha \equiv (1 - k\lambda)(l + m\lambda) \equiv l + (m - lk)\lambda \pmod{\lambda^2}$$

and it suffices to choose k so that $lk \equiv m \pmod{p}$.

A major Hilfsatz (Kummer's lemma) that Kummer will need is:
Let η be a unit of $\mathbb{Q}(\zeta)$ and suppose there is a rational integer a so that

$$\eta \equiv a \pmod{\lambda^p}.$$

Then η is the p th power of a unit of $\mathbb{Q}(\zeta)$.

We can rewrite our equation in terms of ideals as

$$\prod_{r=0}^{p-1}(\alpha + \zeta^r\beta) = (\lambda)^{np}(\gamma)^p.$$

One notices that the ideals on the left do each have a common factor (λ), and that $\lambda \mid (\alpha+\beta)$ implies that $\lambda^2 \mid (\alpha+\beta)$, so $n > 1$. Any other common factor $\mathfrak{d}$, say, must be a common factor of (α) and (β). For the rest, the ideals on the left must be p th powers. So we have

$$(\alpha + \zeta^r\beta) = \mathfrak{d}(\lambda)\mathfrak{a}_r^p \quad \text{and} \quad (\alpha+\beta) = \mathfrak{d}(\lambda)^{p(n-1)+1}\mathfrak{a}_0^p.$$

The point is that the ideals $\mathfrak{a}^p$ may be viewed just as the p th powers of integers of $\mathbb{Q}(\zeta)$. Now consider the identity (somehow known in the transcendence trade as "Siegel's identity")

$$(\zeta - \overline{\zeta})(\alpha+\beta) + (\overline{\zeta} - 1)(\alpha + \zeta\beta) + (1-\zeta)(\alpha + \overline{\zeta}\beta) = 0.$$

After dividing by λ^2 we have integers α'', β', and γ' so that

$$\varepsilon'\lambda^{p(n-1)}\gamma'^p = \beta'^p + \varepsilon''\alpha''^p$$

for units ε' and ε''. But in effect by Fermat's (little) theorem, a p th power of an integer of $\mathbb{Q}(\zeta)$ is a rational integer modulo p, so ε'' is a rational integer modulo p. Thus, and it is here that we invoke Kummer's lemma, ε'' is the p th power of some unit. Hence we have an integer α' in $\mathbb{Q}(\zeta)$ so that

$$\varepsilon'\lambda^{p(n-1)}\gamma'^p = \beta'^p + \alpha'^p,$$

contradicting the minimality of n.

About cyclotomic fields. The class number h of a number field measures the extent to which unique factorization fails; technically, h is the *order* of the *group* of ideal *classes*, that is, of ideals modulo the class of *principal* ideals. A cyclotomic field $\mathbb{Q}(\zeta_p)$ is said to be *regular* if $p \nmid h$.

It has been known for quite a long while (Jensen, 1915) that there are infinitely many irregular primes. Curiously, since heuristically the majority of primes are regular, it is not known whether or not there are infinitely many regular primes.

The class number h^+ of the *real* cyclotomic field $\mathbb{Q}(\zeta_p + \zeta_p^{-1})$ is a divisor of h. It also happens that $h^+ = [E : E_0]$ is the *index* of the subgroup of the units of $\mathbb{Q}(\zeta_p)$ generated by the *real cyclotomic units*

$$\left|\frac{1-\zeta^r}{1-\zeta}\right| = \frac{\sin\frac{r\pi}{p}}{\sin\frac{\pi}{p}} \qquad r = 2, 3, \dots, \tfrac{1}{2}(p-1)$$

in the group E of positive real units of $\mathbb{Q}(\zeta)$.

Kummer showed that if $p \mid h^+$, then $p \mid h_* = h/h^+$. Thus whenever $p \mid h$, certainly $p \mid h_*$. Kummer found a criterion for $p \mid h_*$ that could be checked reasonably conveniently, at least for small p, say less than 100. It is still not known whether in fact ever $p \mid h^+$.

To be precise, Kummer showed that if $p \mid h_*$, then p divides (the numerator of) a Bernoulli number B_{2k} with $2 \le 2k \le p-3$. The irregular primes less than 100 are 37, 59, and 67 (for example, $37 \mid B_{32}$ and $59 \mid B_{44}$).

By 1857 he had completed an analysis of the irregular case to an extent that allowed him to deal with the cases for which p divides at most one of the numbers $B_2, B_4, \ldots, B_{p-3}$, thereby proving the FLT beyond exponent $n = 100$.

I don't see much point here in struggling to explain the calculations which Kummer has to carry out, other than to say that in 1850 Kummer found the criterion that $p \mid h_*$ if and only if there is some integer k with $1 \le k \le (p-3)/2$, so that p^2 divides the sum $\sum_{j=1}^{p-1} j^{2k}$. That may help to explain the appearance of the Bernoulli numbers.

However, by shying away from technicalities and gory details, I do threaten to miss the point of Kummer's contributions to mathematics. Sure, the ancients are a little hard to read, because of poor notation and conventions now strange to us, and because they find some things now easy for us difficult; and, I guess because they gloss over some things we now find difficult. And to top that off, they often fail to write in modern English. It was no great surprise to hear John Coates say, at the time of the Coates–Wiles work on the Birch–Swinnerton-Dyer Conjectures in the complex multiplication case, that core ideas had come from reading Kummer (the context was I think, remarks in praise of Kurt Mahler — who would admonish us to read the masters). Some years later, even I could understand the explanation of how Francisco Thaine's contribution could permit an ingenious transfer of Kummer's ideas from the cyclotomic to the elliptic case. Indeed, Kummer introduced p-adic analysis some 60 years before Hensel.

Specifically, it had been known since 1840 by work of Clausen and of von Staudt that if $k \ge 1$, then $B_{2k} + \sum_{(p-1)\mid 2k} p^{-1}$ is a rational integer. Here and in the sequel p denotes only prime numbers. It follows that if $(p-1) \mid 2k$, then $pB_{2k} \equiv -1 \pmod{p}$. Kummer proves that if $(p-1) \nmid k$, then p does not divide the denominator of B_k/k, and that if further $k \equiv k' \pmod{p^n}$, then $(1-p^{k-1})B_k/k \equiv (1-p^{k'-1})B_{k'}/k' \pmod{p^{n+1}}$. These congruences were just mysterious oddities until Kubota and Leopoldt interpreted them in 1964 in terms of p-adic interpolation of the Riemann ζ-function. And so, on to p-adic L-functions. Thus after all, in the work of Kummer we find the genesis of the eventual solution of Fermat's Last Theorem.

The main material of the Lecture, particularly Kummer's proof, is taken from a lecture I gave in 1977 (I was surprised to rediscover, in my files, the overheads I had used. You may be surprised that I was using prepared overheads at so early a date, but the circumstances were extreme. I was scheduled to speak at a conference at 9am on the morning following the banquet. I thought it likely that neither I nor my audience would be capable of coping with a spontaneous lecture.) I was probably aided by W. J. LeVeque, *Topics in Number Theory*, Vol. II (Reading, Mass.: Addison-Wesley, 1961). I looked at Neal Koblitz, *p-adic Numbers, p-adic Analysis, and Zeta-Functions* , Graduate Texts in Mathematics **58** (New York: Springer-Verlag, 1977), for my concluding remarks.

Notes and Remarks

IV.1 Silly me; I didn't need Sophie Germain. Don Zagier reminds me that FLT I is trivially impossible with $p = 3$ just by considering the matter modulo 9.

IV.2 My summary corrects Kummer. His proof forgets the possibility of a common ideal divisor $\mathfrak{d}$ and omits it entirely.

IV.3 Thaine's work was concerned with bounding the class number of cyclotomic fields, but Karl Rubin realized that "the methods could be applied just as well to anticyclotomic extensions of imaginary quadratic fields, replacing the circular units by elliptic units, thus obtaining his striking results on Tate-Shafarevitch groups of curves with CM." I'm quoting this essentially from p. 16 of Mazur's 'On the passage from local to global in number theory' [*Bull. Amer. Math. Soc.* **29** (1993), pp. 14–50]. Thaine is there quoted explicitly as saying "that at least partial inspiration for [his] ideas came from a close reading of a paper of Kummer".*

IV.4 For Fouvry's work, read Jean-Marc Deshouillers, 'Théorème de Fermat: la contribution de Fouvry' [*Séminaire Bourbaki*, 37ème année, 1984–85, n° 648, juin 1985, in *Astérisque* 133–134 (1986), pp. 309–318]. Fouvry's skills are applauded by Martin Huxley in the infamous limerick

The inversions that Fouvry can do,
Are a blend of the old and the new.
He has shown once again,
That if $q \mid n$,
Then $n = 0 \pmod{q}$.

IV.5 There's plenty of room left for notes so this might be a good spot to interpolate some elementary remarks that really shouldn't be avoided. The first is to say that a very important notion of mathematics is that of *equivalence.*

*Thus I was not surprised to be told by Pieter Moree that similar remarks were made by Wiles in his lectures at Princeton.

For Undergraduates Only

IV.6 Equivalence. It may be contrary to emphasize it, but the notion "equality" in fact serves to *distinguish* things. Each object a belongs to the class C_a containing just a. If objects belong to distinct classes, then they're different, not equal. It's often useful, however, to make do with a "sort of equality", identifying things if their differences are irrelevant for the present purpose. In that case the classes C_a can only distinguish inequivalent elements; each C_a contains all the objects equivalent to a. All this is mathematically interesting if one can, for example, *add* classes. But the trouble in doing so is that $C_a + C_b$ now means the set of all $a' + b'$, with $a' \in C_a$ and $b' \in C_b$. For addition of classes to make sense, for it to be *well defined*, the set of sums has itself to be an equivalence class. In this interesting case, I'll say that the equivalence relation is compatible with, or *respects*, addition. Viewed from the opposite end, one says that addition of the original elements *induces* addition of the classes. The simplest example is congruence modulo m on the integers $\mathbb{Z}$. Here the equivalence classes are the remainder classes $C_0, C_1, \ldots, C_{m-1}$. Usually of course, we denote them just by $0, 1, \ldots, m-1 \pmod{m}$. As this example illustrates, the nub or *kernel* of the matter is the class C_0 containing the neutral element 0. The classes C_a all are given by $C_a = a + C_0$. Because congruence modulo m respects both the operations of $\mathbb{Z}$ it follows easily that its kernel C_0 must be closed under those operations. It has to be a subring of the ring $\mathbb{Z}$. In fact, $C_0 = m\mathbb{Z} = (m)$, the set of all integer multiples of m, is also closed under multiplication by any integer. It's an *ideal* of $\mathbb{Z}$, so-called because it has exactly the properties of an ideal divisor of $\mathbb{Z}$.

IV.7 Morphisms. Mathematics is not concerned with its objects *per se*, it only cares what one can do to them. For example, we learn about vector spaces by studying linear maps — the structure-preserving maps of vector spaces, represented by matrices.

The convenience of a structure respecting equivalence is that as soon as one knows the class C_0 of 0, or the class C_1 of 1 if there is only multiplicative structure, then one knows every class. For C_a must be $C_0 + a = \{x + a : x \in C_0\}$, and in the multiplicative case $C_a = C_1 \cdot a = a \cdot C_1$. But that gives a structure preserving map $a \mapsto C_a$ — such things are called *morphisms* — from the original set to the set of its equivalence classes. The point is that plainly $C_a + C_b = C_{a+b}$ and, if relevant, $C_a \cdot C_b = C_{ab}$. For example, if our relation is congruence modulo m, then the class $C_0 = m\mathbb{Z}$, all integer multiples of m, is the kernel of the story, and we denote the set of equivalence classes — with its arithmetic induced simply by structure preservation — by $\mathbb{Z}/m\mathbb{Z}$. All this would simply be making the simple obscure were it not for the fact that every structure-preserving map of a system S — every morphism of S — is given in this way. For instance, a real vector space V has structure which allows us to add vectors, or to multiply them by scalars. The morphisms f of V are linear maps with domain V; that is, $f(v+v') = f(v)+f(v')$ and $f(rv) = rf(v)$, for $v, v' \in V$ and $r \in \mathbb{R}$. The notion "*linear*" means exactly "structure preserving", and the kernel of f is just the null space of f, all those $v \in V$ which are mapped to the 0-vector. The underlying equivalence relation declares that $v \sim v'$ exactly when $f(v - v') = 0$. We learn that any subspace of V is a kernel of some linear map on V, and conversely every subspace of V determines an equivalence relation and hence a linear map on V.

But it's when we come to algebraic systems a bit more profound than vector spaces that the present viewpoint seems particularly helpful. It is of course dead obvious, if only because $0 + 0 = 0$ or $1 \cdot 1 = 1$, that the kernel equivalence class of a structure respecting equivalence relation must be a subsystem of the domain. Hence subspace, subgroup, subring, sub-gumtree, ...

However, while the kernels of structures respecting some equivalence relations are subsystems, those subsystems often have some additional condition. For a ring the possible kernels are ideals, not just any subring. For a multiplicative group, 1 is the neutral element and C_1 is not just any subgroup, but a *normal* subgroup H with the property that $g^{-1}Hg = H$ for all g in the group. That is, H is closed under *conjugation* by any of the elements g. The fact that in the nonabelian case — the noncommutative case — not every subgroup can be a kernel causes some drastic complications. The general term used for a structure-preserving map is *homo*morphism, to emphasize that structure stays the *same*.

IV.8 Quotient structures. I had better add a warning. To know all about a homomorphism it's not enough just to know what it does to its domain — as fully told by the corresponding equivalence relation. That relation tells us that the image of the map is the system of equivalence classes. One says that it is the original domain *modulo* the equivalence, or — if you wish — *modulo* its kernel. That's the sense in which the remainder classes modulo m are often denoted by $\mathbb{Z}/(m)$. But more, one does have to say where this quotient system lives. If you like, a homomorphism is a quotient map followed by a map embedding the quotient in some codomain. The upshot is that a homomorphism $f : A \to B$ with kernel K corresponds to a sequence

$$0 \longrightarrow K \longrightarrow A \longrightarrow A/K \longrightarrow B\,.$$

Essentially, the morphism f is an onto map $A \to A/K$ followed by a $1:1$ map $A/K \to B$.

If a morphism is $A \to A$, it is called an *endo*morphism. If a homomorphism is both onto and $1:1$ (so that really it only changes the names of things), it's an *iso*morphism. An isomorphism $A \to A$ is an *auto*morphism of A.

IV.9 Fields. If K is a field*, then K contains some smallest subfield, sometimes known as its "prime field". Consider a ring homomorphism $\mathbb{Z} \longrightarrow K$. If its kernel is nontrivial, then it must be a prime ideal (p), because for composite m, the quotient $\mathbb{Z}/(m)$ contains *divisors of zero* — that is, pairs of nonzero elements whose product is 0. That's impossible in a field. On the other hand, it's easy to check, by the remarks in Lectures II and III, that in fact $\mathbb{Z}/(p)$ is the field $\mathbb{F}_p$ of p elements. In this case, when K contains $\mathbb{F}_p$ as its prime field, K is said to have *characteristic* p. If K is finite, then, additively it is an n-dimensional $\mathbb{F}_p$-vector space, whence it contains p^n elements. Its nonzero elements K^* form a cyclic multiplicative group $K^\times$ of $p^n - 1$ elements (because any finite multiplicative subgroup of a field is cyclic). Thus, up to isomorphism, there is exactly one finite field of p^n elements for each prime p and each positive integer n. If the kernel $\mathbb{Z} \to K$ is trivial, that is, if it's just the ideal (0), then the field K must contain all of $\mathbb{Q}$. In this case we say that K has characteristic *zero*.

IV.10 By the way, $\mathbb{F}_p = \mathbb{Z}/(p)$ *is* a field because every nonzero element in it is invertible. This is because an ideal (p) generated by a prime p is *maximal* — every larger ideal, say (p, a), with $p \nmid a$, is all of $\mathbb{Z}$. In particular (p, a) contains 1; that is, there are integers s and t so that $1 = sp + ta$. So $ta \equiv 1 \pmod{p}$, displaying that $t = a^{-1}$ in $\mathbb{F}_p$.

IV.11 Field extensions. We all know the field $\mathbb{C}$ of complex numbers to be just the set $\{a + bi : a, b \in \mathbb{R}\}$; one writes $\mathbb{C} = \mathbb{R}(i)$ to say that $\mathbb{C}$ is the extension by $i = \sqrt{-1}$ of the real numbers $\mathbb{R}$. It comes about like this: The number i is a zero of the irreducible polynomial $X^2 + 1$. One takes the set of all polynomials $\mathbb{R}[X]$ with real coefficients, modulo the ideal $(X^2 + 1)$ of all multiples of $X^2 + 1$ by polynomials. The equivalence relation is tantamount to requiring that X^2 always be replaced by -1. In other words, we might as well admit that $X^2 = -1$; that is, $\mathbb{C}$ is indeed just $\mathbb{R}[i]$, the *polynomials* in i. In a way this is boring, but the point is that *a priori* the extension is $\mathbb{R}(i)$, the set of all *rational functions* in i. However, the ideal (X^2+1) generated by the irreducible polynomial X^2+1 is maximal in $\mathbb{R}[X]$, so automatically the quotient is a field. What's really happening is that the reciprocal of any element $c + di \neq 0$ is itself of the shape $c' + d'i$, thus is a polynomial of degree 1 in i.

Happily, essentially every interesting field extension of any field arises in just this way. Just as I quickly said after Lecture I, if α is the zero of some irreducible polynomial $f(X) \in \mathbb{Q}[X]$ of degree n, then the field $\mathbb{Q}(\alpha)$ is just the polynomials $\mathbb{Q}[X]$ modulo the ideal $(f(X))$. Indeed, $\mathbb{Q}(\alpha)$ is an n-dimensional vector space over $\mathbb{Q}$ — that is, it consists of polynomials with coefficients in $\mathbb{Q}$ and of degree less than n in α, thus it

---

*K for 'Körper' — body. One of the more mysterious questions of mathematics (frankly, quite as intriguing as Fermat's Last Theorem) is why a field should be a body in just about every other language — "corps" in French, "lichaam" in Dutch, ... However, Igor Shparlinski reminds me that it is "поле" — field — in Russian.

is an n-dimensional $\mathbb{Q}$-vector space with basis $\{1, \alpha, \dots, \alpha^{n-1}\}$ — just as, analogously, $\mathbb{C}$ is a two-dimensional $\mathbb{R}$-vector space.

Remarkably, it is true that a sequence of such finite extensions of $\mathbb{Q}$, for example $\mathbb{Q}(\sqrt{2})(\sqrt{3})$, is always generated by a single element. This is sometimes called the *theorem of the primitive element*; the example is $\mathbb{Q}(\sqrt{2} + \sqrt{3})$.

Mind you, these remarks hold for "finite" extensions. Nothing of this sort is true for infinite extensions such as $\mathbb{A} = \overline{\mathbb{Q}}$, the field of all algebraic numbers. One refers to a *finite* extension of $\mathbb{Q}$ — thus, a field $\mathbb{Q}(\alpha)$ — as an *algebraic number field*, or just a *number field*.

IV.12 Galois groups. I've given both concrete and abstract definitions of a number field. Yet these definitions are not quite the same. On the one hand, $\mathbb{Q}(\sqrt[3]{2})$ is the set $\{a + b\sqrt[3]{2} + c\sqrt[3]{2^2} : a, b, c \in \mathbb{Q}\}$; on the other hand, more abstractly, it's the set of equivalence classes $\mathbb{Q}[X]/(X^3 - 2)$. Yet the latter just as much gives the field $\mathbb{Q}(\omega\sqrt[3]{2})$, with ω a complex cube root of 1. This ambiguity reminds one to talk of such *conjugate* fields all at once so as not to show prejudice in favor of one rather than the other. This means speaking all at once about all the fields $\mathbb{Q}(\sigma\alpha)$, where $\sigma\alpha$ runs through all the zeros of the defining polynomial of α over $\mathbb{Q}$. Better, one thinks of σ as a map permuting those different zeros. The σ's form the *galois group* $G(\mathbb{Q}(\alpha)/\mathbb{Q})$ of the field extension $\mathbb{Q}(\alpha)$ of the rationals $\mathbb{Q}$. Any such group is a subgroup of the group $G(\overline{\mathbb{Q}}/\mathbb{Q})$ of all structure-preserving permutations of the roots of any irreducible equation over $\mathbb{Q}$.

IV.13 Galois theory. The correspondence between field extensions and galois groups was first noticed in the endeavor to explain why there should be a "formula" for finding the roots of polynomial equations ofdegree less than 5, but apparently no such formula for degree 5 or more. Now the galois group of the "general" equation of degree n is the group S_n, of all permutations of its n roots. And indeed, the cases can then be seen to be different. Each S_n has a subgroup $\mathcal{A}_n$ of all even permutations, of index 2 in S_n. However, for $n \geq 5$ the $\mathcal{A}_n$ are *simple* — having no proper normal subgroup. For small n there are descending chains of normal subgroups $S_4 \supset_2 \mathcal{A}_4 \supset_3 V \supset_2 S_2 \supset_2 \{1\}$; $S_3 \supset_2 \mathcal{A}_3 \supset_3 \{1\}$; and $S_2 \supset_2 \{1\}$. Here the index of each normal subgroup is marked as a subscript on the superset (containment) sign. That all such indices are primes p shows that all roots of equations of low degree can be obtained by appending a sequence of radicals $\sqrt[p]{a}$. The impossibility of a "formula" for the general equation of degree 5 is, naturally, another topic in *Abstract Algebra and Famous Impossibilities* by Arthur Jones, Sydney A. Morris, and Kenneth R. Pearson (New York: Springer-Verlag, 1991).

Pierre de Fermat

engraving by F. Poilly
Varia Opera Mathematica D. Petri de Fermat ... , Tolosæ 1679.

Notes on Fermat's Last Theorem

LECTURE V

By the powers conferred on us, by Dr. Paul Wolfskehl, deceased of Darmstadt, hereby we fund a prize of one hundred thousand Marks, to be given to the person who will be first to prove the Great theorem of Fermat.

Göttingen, June 27, 1908
Die Königliche Gesellschaft der Wissenschaften

In the first year of the Wolfskehl Prize, 621 "solutions" were submitted. That surely raises the question of how, and why, do mathematicians make mistakes? Of course, I don't want to philosophize that humans are born to sin and error (though the trigonometric allusion deserves comment). Nor do I really want to comment on blunders, slips of the pen, or misprints. Moreover, no doubt the 621 "solutions", and the thousands that followed, contained a great deal of outright nonsense. By that I intend to include pure principle, say that of maintaining that it is God's will that indeed a square splits into two squares, but therefore "of course" that a cube split into at least *three* cubes, a fourth power into at least *four* fourth powers, and so on. Euler is sometimes blamed for a conjecture to that effect, but whatever, it's outright wrong. One of the few times that I've been able to send an amateur scurrying away was by pointing out that sadly,

$$144^5 = 27^5 + 84^5 + 110^5 + 133^5 .$$

Of course, slips of the pen can be followed by an entire coherent chain of correct reasoning and may be astonishingly hard to pick. If one has a great deal invested in an argument, one may be emotionally incapable of being sufficiently critical. It helps to be aware of the principle, once nicely put to me by Kurt Mahler, according to which one cannot expect to get *gehaktes Rindfleisch* from a meatgrinder unless one has put meat into that grinder. In other words if I prove some surprising fact, I had better have had some pretty clever idea, or have used some wondrous secret. On occasions that I have looked at questions the great ones couldn't handle I have always asked myself what new knowledge, or new combination of old knowledge, I had available; in other words, what *meat* did I have?

The best way to find absurd errors or unjustified assumptions is to have a student read your manuscript. The impact of an innocent question such as "I couldn't quite understand why it's right to write 1 = 2" can be shattering. It is also bad for one's self-confidence to be asked to explain why it is correct to neglect a blatant counterexample. I recall a report on a survey which asked mathematicians how many papers they had read in complete detail the previous year. It appears that the average paper is read by some 0.76 mathematician, including author, referee, and reviewers. It is little wonder that actually having one's paper read can be revealing.*

We also make mistakes by our use of "obviously", "evidently", "clearly" and the like. Now it is the case that no argument tries to dot every "t" and cross every "i". Even correct arguments will contain such "gaps". But the trouble is that we may have conquered the apparent obstruction without realizing that we have now brought a real difficulty to light. It's that kind of thing that causes such admissions as "my proof developed a gap, and that gap became a chasm that swallowed the proof". Such situations can lead to premature announcements. Nonetheless, in that the real problem may have been revealed, success may follow after all. One conquers and then, slowly and reluctantly, realizes that all that has been done thus far is to expose a true obstruction to a proof. Sometimes moreover, the sheer excitement of it all seems to allow one to leap the new abyss, thereby doing something far more clever than one is normally capable of. Suddenly one has proved Theorem 3 (van der Poorten), or better yet, Theorem B. "Amateurs" don't know the principle of the meatgrinder. They believe that jumping a crack will suffice.

I began this tirade in an attempt to explain why we believe that Andrew Wiles indeed does have a proof of the FLT. Wiles presented his work at the workshop (June 1993) on "Iwasawa theory, automorphic forms and p-adic representations" at the Isaac Newton Institute for Mathematical Sciences in Cambridge, England in a sequence of three lectures with title "Elliptic curves, modular forms, and Galois representations". Rumors had circulated for some days that something interesting was about to be said and excitement mounted as it became apparent that Wiles might be able to claim a fundamental advance in the matter of Taniyama's conjecture. His expert audience saw quality meat. They saw new combinations of ideas that might allow chasms to be jumped. The subject is understood well enough for it to be very surprising indeed if real problems after all reside in the technicalities, necessarily slurred in the course of Wiles' presentation. Thus there is real confidence in Wiles' corollary: *Suppose that p, u, v, and w are integers, with $p > 1$. If $u^p + v^p + w^p = 0$ then $uvw = 0$.* We will not be needing to use the fine remark of Allan Adler:

> *... thus, the entire proof of Fermat's Last Theorem collapses like a house of cards. The great problem is still unsolved and they are right in the Star Trek*

---

*Peter Merrotsy reminds me of a dictum that one's Ph.D. thesis should be readable by at least two people, one of them by preference being the author.

episodes[†] *when they say that in the twentythird century the problem is still open!*

With that off my chest let me say a few additional things about the work on the FLT following Kummer. I became seriously interested in Fermat's Last Theorem in the seventies, partly because of some related work I had done[†]. Curiously, I also wrote about some calculations on the Taniyama-Weil Conjecture at much that time*, of course with no idea that it might be related. I was surprised to find how little of great substance had been achieved on the FLT in the previous fifty years. Sure there was the very extensive work of Vandiver, filling minor gaps in Kummer's work and simplifying his criteria, and substantial generalizations of Mirimanoff's congruence conditions in the first case. Sadly some of those generalizations were disputed, as much as anything else I suspect, because barely anyone was prepared to plough through the details of the papers. But perhaps I should not deride this century's work. You can make your own judgment after reading Paulo Ribenboim's *13 Lectures on Fermat's Last Theorem* (New York: Springer-Verlag, 1979).

Of course, the arrival of the computer made it feasible to carry out very extended calculations. In that way, the first case was verified up to the bit length, 3×10^9 or so [J. Brillhart, J. Tonascia and P. Weinberger, 'On the Fermat quotient', in O. Atkin and B. J. Birch, *Computers in Number Theory*, Academic Press, 1971, pp. 213–222], and the general case up to 125 000 [S. S. Wagstaff, 'The irregular primes up to 125,000', *Math. Comp.* **32** (1978), pp. 583-591].

However a curious exception to what I saw as a desert was an observation of Terjanian, who showed that if the equation $x^{2p} + y^{2p} = z^{2p}$ has a solution, then $2p$ divides one of x or y [G. Terjanian, 'Sur l'équation $x^{2p} + y^{2p} = z^{2p}$', *C. R. Acad. Sc. Paris*, **285** (1977), pp. 973–975]; here p is an odd prime. This settled the first case for even exponents. The argument is entirely elementary. We already know that one of x and y, say x, must be even and, by Abel's formulas, that

$$\gcd(z^2 - y^2, (z^{2p} - y^{2p})/(z^2 - y^2))$$

[†]Gary Cornell tells me that in the episode "The Royale", Captain Picard purports to relax by studying the FLT.

[†]'Some remarks on Fermat's conjecture' (with K. Inkeri), *Acta Arith.* **XXXVI** (1980), pp. 107–111.

*'The polynomial $X^3 + X^2 + X - 1$ and elliptic curves of conductor 11', *Séminaire Delange-Pisot-Poitou (Théorie des nombres)* (18e année: 1976/1977), Fasc.1, exp. n°17, 7pp (Secrétariat Mathématique, Paris, 1977) and 'Elliptic curves of conductor 11' (with M. K. Agrawal, J. H. Coates and D. C. Hunt), *Math. Comp.* **34** (1980), pp. 991-1002.

The dates now surprise me and remind me of the extraordinary delays then current in publication. Some years later John Loxton and I accidentally dedicated an article to the 80th birthday of Paul Erdős (rather than the 75th), leading Andrzej Schinzel to say that he accepted the paper subject to one change, unless we wanted it kept for five years. I was able to retort that my error was understandable given the way that Erdős carries on about his age — smile from Schinzel — and that anyhow given *Acta Arithmetica* delays, it was probably spot on — laughter from everyone else.

is 1 or p. So it suffices to show that the gcd cannot be 1. But if the two factors are relatively prime, both must be squares. However, by playing with quadratic reciprocity, Terjanian shows by induction on the integers m and n when y and z are odd that $(z^{2m} - y^{2m})/(z^2 - y^2)$ is a square modulo $(z^{2n} - y^{2n})/(z^2 - y^2)$ if and only if m is a square modulo n. That suffices. The fact that Terjanian's result was not known until then, and the elementary nature of the argument, provided a great surprise.

In 1974, Rob Tijdeman demonstrated the power of Baker's method by dealing with Catalan's conjecture to the effect that the only pair of "perfect powers" differing by 1 are $3^2 = 9$ and $2^3 = 8$. The equation $x^u - y^v = 1$ in four integer variables, all at least 2, might seem more complicated than Fermat's Last Theorem. It was a surprise that his refinement of Baker's method allowed Tijdeman to show that all four variables are bounded by effectively computable constants. However, it remains computationally infeasible to show that indeed $3^2 - 2^3 = 1$ displays the only solution. Nor have we any handle at all, as yet, on the conjecture that for any $z \neq \pm 1$, fixed, or varying with just given prime factors, the equation $x^u - y^v = z$ also has essentially only a few small solutions. It was this problem that led to my interest in the FLT.

In 1978 I had several months of pleasure in having used Baker's method to prove, subject to Vandiver's conjecture that p does not divide h^+, that Fermat's Last Theorem was true for all sufficiently large exponents. My argument survived the one-week and one-month tests, but my suspicions became certainty when I found that my arguments were so powerful that they handled the FLT without hypothesis and could even prove some false "facts". To find my blunder I ultimately had to resort to the well-tried method of throwing the manuscript into the air and studying the page at which it opened. So it goes. My interest in the FLT faded and I have been forced to a study of *Mathematical Reviews* — fortunately now available on CD — to check on the years until now.

Andrew Granville showed that 'The first case of Fermat's Last Theorem is true for all prime exponents up to 714, 591, 416, 091, 389' [*Trans. Amer. Math. Soc.* **306** (1988), pp. 329–359], *inter alia* by checking and extending the consequences of the Kummer-Mirimanoff congruences and proving that a first case solution having exponent p entails that $p^2 \mid (r^p - r)$ for all primes $r \leq 89$. In 1992, J. P. Buhler, R. E. Crandall and R. W. Sompolski used fast Fourier transforms and more efficient Bernoulli number identities to determine 'The irregular primes to one million' [*Math. Comp.* **59** (1992), pp. 717–722] and verified Vandiver's conjecture that $p \nmid h^+$, and Fermat's Last Theorem, for those primes. Indeed, those results were then extended in J. P. Buhler, R. E. Crandall, R. W. Sompolski and T. Metsänkylä, 'Irregular primes and cyclotomic invariants to four million' [*Math. Comp.* **61** (1993), pp. 151–153].

Yet more dramatic results appeared variously as corollaries of Faltings' 1983 proof of Mordell's Conjecture*. Faltings' work entails, as a very particular case, that for each exponent $n \geq 4$ the equation $x^n + y^n = z^n$ has at most finitely many solutions. Among others, Granville and Heath-Brown remarked that it is a relatively "cheap" corollary that the FLT holds for almost all exponents n, in the sense that the probability of any n providing a counterexample is zero, notwithstanding our still not knowing that it holds for infinitely many *prime* exponents.

Relying on a result of Etienne Fouvry ['Théorème de Brun-Titchmarsh: application au théorème de Fermat', *Invent. Math.* **79** (1985), pp. 383-407], Len Adleman and "Roger" Heath-Brown [*Invent. Math.* **79** (1985), pp. 409-416] showed that the first case of the FLT holds for infinitely many prime exponents, the first result of that breadth. I quote a review of the late Emil Grosswald (MR: 87d:11020) in saying that

> ... the tools used, or quoted, in the proof are rather formidable and comprise, among others, a [very particular case of a recent] theorem of Faltings; old theorems of Sophie Germain and of Wieferich and Mirimanoff and a generalization of Sophie Germain's theorem; the Siegel-Walfish-Page theorem on primes in arithmetic progressions; the Bombieri-Vinogradov prime number theorem; Linnik's theorem (smallest prime in an arithmetic progression); the Brun-Titchmarsh theorem; Chebyshev's lower bounds estimates obtained from upper bounds; sieve methods, especially Rosser's sieve; Kloosterman sums and their evaluations by Weil, Iwaniec, and Kuznetsov; modular functions and the non-Euclidean Laplacian operator, etc.

The array of notions that we would need mention to detail Wiles' eventual proof of Fermat's Last Theorem is yet more impressive — or depressing, for those who retained the dream of an elementary proof. For the rest, in *Mathematical Reviews* one finds mostly minor comments. There is also a note of Francisco Thaine ['On the first case of Fermat's last theorem', *J. Number Theory* **20** (1985), pp. 128-142], whose useful remarks on the Kummer-Mirimanoff congruences eventually affect our understanding of elliptic curves.

As remarked, by 1993 the FLT was known for all exponents to 4 million or so. Yet, had it not been for the recent Frey-Ribet observations showing that the truth of the FLT followed from the Taniyama-Weil Conjecture, one could well have repeated Harold Edwards' remark in the introduction to his book [*Fermat's Last Theorem*, Graduate Texts in Mathematics **50** (New York: Springer-Verlag, 1977)] that "there is as yet little reason to believe the Theorem".

I found the opening quote and remark in Paulo Ribenboim, *13 Lectures on Fermat's Last Theorem*, (New York: Springer-Verlag, 1979). The felicitous exclamation by Allan Adler

* See Lecture VI. Faltings proved that curves of genus at least 2 contain at most finitely many rational points. If n is 4 or more then the Fermat equations define such curves.

is from his FLT notes, once available on the e-math gopher server. It arises in the context of his explanation of the notion 'minimal model'.

Notes and Remarks

V.1 To err is human, to moo bovine.

V.2 When I gave this lecture it wasn't clear to me that Euler should be blamed for a conjecture to the effect that if $n \geq 4$, then an equation

$$x_1^n + x_2^n + \cdots + x_{n-1}^n = x_n^n$$

has no solutions in positive integers. However, had I checked Dickson's *History of the Theory of Numbers*, Vol. II: *Diophantine Analysis* (New York: Chelsea, 1971), p. 648*ff* I would have seen that Euler did claim *circa* 1769 that such a theorem is "hardly to be doubted" for $n = 4$ and that "in the same manner it would seem to be impossible to exhibit four fifth powers whose sum is a fifth power, and similarly for higher powers". The counterexample I quote for $n = 5$ was obtained by computer search by L. J. Lander and T. R. Parkin ['Counterexamples to Euler's conjecture on sums of like powers', *Bull. Amer. Math. Soc.* **72** (1966), p. 1079]. It was the only one known until recently.

Apparently I haven't had to deal with difficult amateurs for a while, for if I had, I should have offered the more recent and yet more dramatic

$$422481^4 = 95800^4 + 217519^4 + 414560^4,$$

which is known to be the smallest counterexample to Euler's conjecture for fourth powers. Noam Elkies, 'On $A^4 + B^4 + C^4 = D^4$' [*Math. Comp.* **51** (1988), pp. 825–835] caused considerable surprise by showing that there are infinitely many counterexamples to Euler's conjecture for $n = 4$. He cites

$$2682440^4 + 15365639^4 + 18796760^4 = 20615673^4$$

as well as the minimal example, which was obtained subsequently using a computer search.

V.3 Everyone knows that it's easy to express a square as a sum of two squares. We all begin by quoting $5^2 = 4^2 + 3^2$. It's just as easy to give an example of a cube expressed as a sum of three cubes. For indeed, simply $6^3 = 5^3 + 4^3 + 3^3$.

V.4 Concerning the so-called Fermat quotients $q_p(2) = (2^p - 2)/p$, I recently (May 1995) heard Karl Dilcher describe work of himself, Crandall, and Pomerance showing that the only primes less than 4×10^{12} for which $q_p(2) \equiv 0 \pmod{p}$, remain just 1093 and 3511. Nonetheless, everyone continues to believe that there are infinitely many prime numbers p so that $q_p(2) \equiv 0 \pmod{p}$. The point is that the Fermat quotients appear to behave randomly modulo p, so the probability that $q_p(2)$ is divisible by p is seemingly $1/p$. But the sum $\sum_p 1/p$ taken over all the primes p diverges and so the probability that some additional $q_p(2)$ is divisible by p readily exceeds certainty, to wit 1. Proving such a thing, however, remains totally out of reach.

V.5 Mind you, notwithstanding my claims that it is reasonable to believe that Wiles has a proof, one should expect the odd alarum. Thus, for example, in mid-October (1993) I saw the following message: "There appears to be a gap in Andrew Wiles' strengthening of Matthias Flach's work on the symmetric square, so that it seems that Fermat is not quite proven yet. There are subtle questions involving a trivial zero which he now cannot seem to avoid. However, I am convinced that the desired result about the symmetric square can be established soon by the methods we have in hand" (John Coates, 19 October, 1993).

Subsequently, the world was awash with rumors that a serious gap had developed. These included, according to my informant Mark Sheingorn, a long article in the *Newark Star-Ledger* entitled "Waiting for Fermat", saying that there is a gap, that the experts think it will take Wiles two months to a year to fix it, and thus, ... Apparently, Enrico Bombieri was quoted to explain what a "gap" is ... .

V.6 But then matters seemed to worsen. On 19 November John Coates gave a talk, at the Newton Institute, Cambridge, on Fermat's Last Theorem in which he said that the experts had now looked at Andrew Wiles' paper and there is a small hole in the proof which is not obviously fixable. The argument involves some version of a conjecture of Tamagawa that Wiles also needed in his proof. Coates further said that there is one promising line of argument for fixing the hole, but it may take a year or two ... . The main message was: "If you managed to get a T-shirt, hold on to it, they may become collector's items yet!" [This alludes to the white T-shirt announcing the proof at Cambridge University, June 1993. At the *Cursos de Verano* (Summer Course) on Fermat's Last Theorem at El Escorial in August 1994, Cassels wore one through the entire week.]

V.7 The authoritative words on the matter came in early December 1993 when Andrew Wiles issued the following statement to the sci.math newsgroup on the internet:

> In view of the speculation on the status of my work on the Taniyama-Shimura Conjecture and Fermat's Last Theorem I will give a brief account of the situation. During the review process a number of problems emerged, most of which have been resolved, but one in particular I have not yet settled. The key reduction of (most cases of) the Taniyama-Shimura Conjecture to the calculation of the Selmer group is correct. However the final calculation of a precise upper bound for the Selmer group in the semistable case (of the symmetric

square representation associated to a modular form) is not yet complete as it stands. I believe that I will be able to finish this in the near future using the ideas explained in my Cambridge lectures.

The fact that a lot of work remains to be done on the manuscript makes it still unsuitable for release as a preprint. In my course in Princeton beginning in February I will give a full account of this work.

V.8 On December 13, Mark Sheingorn e-wrote me (and others): "I assume we all saw today's article on page 9 of the *Times*. It is basically pro-Wiles, relegating his handling of the MS until the last. It says the major problem is computing the size of the Selmer group ... , his major technical feat. Ribet says his inequality is 'very, very likely to be true', but there is no similar endorsement of the extension of Flach's Euler system argument to obtain the inequality. Still, this is only the *Times*, but they did have the symmetric square statement correct. J. J. Kohn has a deft negative: 'It would be really premature to say he doesn't have a proof'."

V.9 A week later I was in Hong Kong at a conference on "Elliptic curves and modular functions" starring most of the principal players. While there I was reminded just how substantial a contribution Wiles has made to our dealing with the Taniyama-Shimura-Weil Conjecture. On returning home, over Christmas I wrote a diatribe for consumption by the real world. It appears as "Remarks on Fermat's Last Theorem", an appendix to this book.

V.10 At the very beginning of April 1994, Henri Darmon e-perpetrated the following announcement:

> There has been a really amazing development today on Fermat's Last Theorem. Noam Elkies has announced a counterexample, so that FLT is not true after all! He spoke about this at the Institute today. The solution to Fermat that he constructs involves an incredibly large prime exponent (larger than 10^{20}), but it is constructive. The main idea seems to be a kind of Heegner-point construction, combined with a really ingenious descent for passing from the modular curves to the Fermat curve. The really difficult part of the argument seems to be to show that the field of definition of the solution (which, a priori, is some ring class field of an imaginary quadratic field) actually descends to $\mathbb{Q}$. I wasn't able to get all the details, which were quite intricate ... .
>
> So it seems that the Shimura-Taniyama Conjecture is not true after all. The experts think that it can still be salvaged, by extending the concept of automorphic representation, and introducing a notion of "anomalous curves" that would still give rise to a "quasi-automorphic representation".

I noted that the actual construction of his anomalous solution would have been quite some feat on Elkies' part,* given that such a solution would, according to Inkeri's results, necessarily involve numbers at least some 10^{21} digits long.

Noting the date of the announcement, I sarcastically replied to my New York informant that "Personally, I remain a little suspicious of strange allegations made on Good Friday." One need not go searching for an old diary or calendar to check the date of Fridays in 1994. An algorithm of John Conway [described in an article of mine, 'A dozen years is but a day', *Austral. Math. Soc. Gazette* **11** (1984), pp. 33–34] allows one to deduce the matter by pure thought.

V.11 It would have been dissatisfying to have had to finish this book with Wiles' proof still incomplete. Fortunately, I suffered writer's block for a sufficiently long time for Wiles to settle the matter. On 25 October 1994 Karl Rubin made the following announcement:

> As of this morning, two manuscripts have been released
>
> 'Modular elliptic curves and Fermat's Last Theorem',
> by Andrew Wiles
> 'Ring theoretic properties of certain Hecke algebras',
> by Richard Taylor and Andrew Wiles.
>
> The first one (long) announces a proof of, among other things, Fermat's Last Theorem, relying on the second one (short) for one crucial step.
>
> As most of you know, the argument described by Wiles in his Cambridge lectures turned out to have a serious gap, namely the construction of an Euler system. After trying unsuccessfully to repair that construction, Wiles went back to a different approach, which he had tried earlier but abandoned in favor of the Euler system idea. He was able to complete his proof, under the hypothesis that certain Hecke algebras are local complete intersections. This and the rest of the ideas described in Wiles' Cambridge lectures are written up in the first manuscript. Jointly, Taylor and Wiles establish the necessary property of the Hecke algebras in the second paper.
>
> The overall outline of the argument is similar to the one Wiles described in Cambridge. The new approach turns out to be significantly simpler and shorter than the original one, because of the removal of the Euler system. (In fact, after seeing these manuscripts Faltings has apparently come up with a further significant simplification of that part of the argument.)

---

*Incidentally, Elkies was just an innocent bystander in this affair, with his prowess borrowed to lend credence to the claims.

V.12 In effect, Flach's Euler systems had given Wiles the confidence to complete and present his arguments. In the event, the idea could not be sustained, but a return to Plan A did the trick after all. Almost all research mathematicians will confess to analogous experiences. One has long worked on a problem. Then some marvelous new insight leads one willy-nilly to complete the work. All too often, at the end of the day, that breakthrough insight has turned out to be nonsense. It plays no role in the final argument other than to have made it psychologically possible.

V.13 At this point one could safely report, as did Fernando Gouvêa, that

> *A reckless young fellow from Burma,*
> *Found proofs of the theorem of Fermat.*
> *He lived then in terror,*
> *Of finding an error,*
> *Wiles's proof, he suspected, was firmer.*

V.14 It's nice to hear from one's friends, so I was indeed amused to have Bill McCallum write me in December 1994 that

> ... in preparation for a talk on Fermat's Last Theorem to undergraduates, I downloaded your lecture notes, where I read about your evanescent proof that Vandiver's conjecture implies FLT for sufficiently large exponents in the late 70s. You might be amused to know that recently I worked out how to prove that Vandiver's conjecture implies FLT using the methods I started developing while I was at Macquarie. I was in the process of writing this up when I heard the news of Andrew's second announcement. C'est la vie!

In fact, Bill did subsequently announce a proof that Vandiver's conjecture, that l does not divide h_l^+, does imply Fermat's Last Theorem, using "a striking application of the method of Chabauty and Coleman". He had earlier applied those ideas to the second case of the FLT for regular primes in W. G. McCallum, 'On the method of Chabauty and Coleman' [*Math. Ann.* **299** (1994), pp. 565–596]. Bill's work too suffered severe alarums and gaps and, notwithstanding continuing confidence in the method, its claims had been formally withdrawn at the time of final editing of this book.

V.15 The remark alluded to at the end of the lecture is this: Edwards observed that since no sufficient condition for Fermat's Last Theorem had ever been shown to include an infinite number of prime exponents, "one is in a position to prove Fermat's Last Theorem for virtually any prime within computational range but one cannot rule out the possibility that the theorem is false for all primes beyond some large bound".

Notes on Fermat's Last Theorem

LECTURE VI

Our only certain knowledge of Diophantus rests upon the fact that he quotes Hypsicles [~ −150] *and that he is quoted by Theon Alexandrinus* [whose date is fixed by the solar eclipse of June 16, 364].

O. Neugebauer, *The Exact Sciences in Antiquity*

Number theory was strong in antiquity. But the books of Diophantus were lost in the burning of the library of Alexandria and had little influence on mathematics until the seventeenth century, when Fermat was inspired by Bachet's recent translation.* The ideas underlying the solutions to the problems in the *Arithmetica* were substantially in advance of those then current in the West.

Diophantus is largely concerned with the problem of finding a rational solution to various equations; we recognize the methods are geometrical. Dealing with Pythagorean triples, he considers the equation $x^2 + y^2 = 1$ in rationals x and y; which we well know to be a circle. An obvious, albeit trivial, point on this locus is $(-1, 0)$. A typical line through that point is parametrized by $x = u - 1$, $y = tu$. This line intersects the circle when $u^2 - 2u + 1 + t^2u^2 = 1$; and happily we can cancel the known solution $u = 0$ to obtain $u = 2/(1 + t^2)$, yielding a new point $x = (1 - t^2)/(1 + t^2)$, $y = 2t/(1 + t^2)$. This is of course essentially the solution of Lecture I, now illustrating that the circle may be parametrized by rational functions. In this case we get infinitely many solutions, given by a simple formula. Different problems might have infinitely many solutions not given by a rational formula, or just finitely many solutions, or none at all.

Problem 24 of Book IV of Diophantus suggests that we split a given number, say 6, into two parts so that their product is a cube minus its cube root. That is, $y(6 - y) = x^3 - x$. Once again, $(-1, 0)$ provides a trivial solution but now when we try $x = u - 1$, $y = tu$ we get, after canceling the known solution $u = 0$,

$$t(6 - tu) = (u - 1)(u - 2) \quad \text{or} \quad u^2 - (3 - t^2)u + (2 - 6t) = 0.$$

*Six of the thirteen books came into the hands of the astronomer Johannes Müller (*Regiomantus*) in 1464; quite recently four additional books came to light; see J. Sesiano, *Books IV to VII of Diophantus' Arithmetica*, (New York: Springer-Verlag, 1982).

In general, this leads to irrational values for u. However, on selecting the slope $t = \frac{1}{3}$, we may cancel once more to obtain $u = \frac{26}{9}$ whence $x = \frac{17}{9}$, $y = \frac{26}{27}$ is a new solution. The fun thing is that we can now construct the tangent at this new solution to find a further solution, and so on. In general, given two solutions, the secant yields a third solution. Of course the complexity of the solutions threatens to increase dramatically.

For example, writing to Mersenne in 1643, Fermat asks for right-angled triangles so that both the hypotenuse and the sum of the two sides are squares. Taking the sides as $\frac{1}{2}(1 \pm y)$ and the hypotenuse as x^2, we of course have $y^2 = 2x^4 - 1$ by Pythagoras. Fermat can compute the basic solution $P(13, 239)$. But the geometrical problem requires that the sides of a triangle be positive numbers, so $-1 < y < 1$. The tangent at P provides a further solution $2P(\frac{1525}{1343}, \frac{2750257}{1803649})$. This still won't do. Finally, we get $3P(\frac{2165017}{2372159}, \frac{3503833734241}{5627138321281})$ from the secant through two solutions, corresponding to a triangle with sides:

$$a = 1061652293520, \quad b = 4565486027761, \quad c = 4687298610289.$$

Not only is this *a* solution, but the method guarantees that this is the *smallest* solution!

With yet higher-degree equations these methods fail in general. The upshot is that given a polynomial equation $f(x, y) = 0$ with integer coefficients, there are three cases, seemingly depending on the (total) degree of f. Namely, if f is of degree at most two, we have none or infinitely many solutions — these cases are parametrized by rational functions. This is the case of *rational curves* — curves of genus 0. If f is of degree 3 we may have finitely many — the method of *infinite ascent* of the last examples may cycle, or infinitely many solutions. This is the case of *elliptic curves*; that is, curves of genus 1. These curves do not, of course have anything to do with ellipses — those are conics and may be parametrized by rational functions. Rather, the point is that these curves are parametrized by the so-called elliptic functions*. Finally, there are curves of *general type*, of genus $g \geq 2$, which seem only to have sporadic rational points.

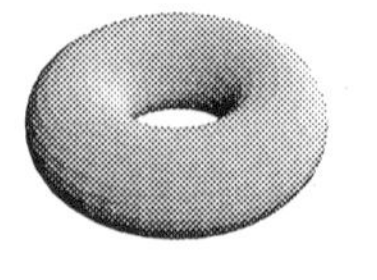

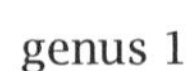

genus 1 genus 2

The notion of *genus* arises by considering the set of complex points (z, w) so that $f(z, w) = 0$. One now has a *surface* of solutions, whose

*Which I'll talk about in my next lecture.

closure is topologically a sphere in the rational case, a torus in the elliptic case ("a sphere with one hole"), and a surface with more than one hole in the general case. These notions go back to Riemann — hence *Riemann surface*, and were studied by Poincaré at the turn of the century.

In any event, the elliptic case provides the interesting collection of diophantine problems. Siegel did prove in 1929 that there are only finitely many *integer* points on curves of genus one or more, but in 1983 — one of the major results of this century — Faltings settled Mordell's Conjecture to the effect that curves with genus greater than one have only finitely many *rational* points. Thus, since then we have known by that result that any one Fermat equation $x^n + y^n = z^n$ with $n \geq 4$ can have at most finitely many solutions.

In the elliptic case the solutions have a structure revealed by the method sketched above. One can always *model* elliptic curves by an equation $y^2 = x^3 + ax + b$ with $4a^3 + 27b^2 \neq 0$ — this may involve strategically locating points at infinity to reduce the degree from 4 to 3. For example, Fermat's equation $u^3 + v^3 = w^3$ corresponds to $y^2 = x^3 - 432$ by the transformation $x = 12w/(u+v)$ and $y = 36(u-v)/(u+v)$; the elliptic curve $y^2 = 2x^4 - 1$ requires more work than I am prepared to invest at this moment. Given a cubic model for the curve, a straight line cuts the curve at three points P, Q, and R, say, and we may write $P + Q + R = 0$, with 0 being the point with $y = \infty$. If $P = (x, y)$, then $-P = (x, -y)$. A tangent at P, which "cuts" the curve twice at P, also cuts the curve at $-Q$, that is, so that $2P + Q = 0$. Obviously, all this defines a commutative operation, but, remarkably this "addition" is also associative — as one may show using the fact that three cubic curves with eight points in common also have a ninth common point.

Thus the rational points on an elliptic curve form an abelian group. More than that, in 1922 Mordell confirmed a suggestion of Poincaré that the group is *finitely generated*, here meaning that one need start with just finitely many points to generate all the rational points by the construction described above. Finitely generated abelian groups are all of the shape $\mathbb{Z}^r \oplus T$, that is, they have a finite *torsion* part T consisting of points of *finite* order — those points P for which $mP = 0$ for some m — and an infinite part $\mathbb{Z}^r$. So, deciding whether there are infinitely many rational points on an elliptic curve reduces to determining whether its *rank* r is or is not 0. We now know the possibilities for the torsion subgroup. In 1977, Barry Mazur established that if the cubic $x^3 + ax + b$ is irreducible, then $T \cong C_{2n+1}$, a cyclic group of order $2n + 1$, with $n \leq 4$; if the cubic has just one rational linear factor, then $T \cong C_{2n}$ with $n \leq 6$; and if the cubic splits completely, then $T \cong C_2 \times C_{2n}$ with $n \leq 4$. Whatever, it is not too difficult to find all the torsion points in any particular example. To be precise, presuming that the parameters a and b are integers, a fairly elementary theorem of Nagell and Lutz entails that a torsion point has integral coordinates and that the y coordinate is either 0 (in which case $2P = 0$) or it is a divisor of $4a^3 + 27b^2$. Moreover, it is also elementary to argue that $E(\mathbb{Q})$ is a subgroup of the group $E(\mathbb{R})$ of real points on the curve, and that $E(\mathbb{R})$ is isomorphic to the circle $\mathbb{R}/\mathbb{Z}$ or to the product $\mathbb{R}/\mathbb{Z} \times \mathbb{Z}/2\mathbb{Z}$, according as $4a^3 + 27b^2$ is positive or negative. In this way one sees that the content of Mazur's very deep result is the various bounds on n, yielding just 15 cases for rational torsion.

In 1928 André Weil generalized Mordell's Theorem to elliptic curves over number fields (and in 1940 to the case of abelian varieties). Thereby we now speak of the group $E(\mathbb{Q})$, of rational points on an elliptic curve E, as the *Mordell-Weil group* of the curve.

Let me now return to a classical problem to illustrate the difficulties in guessing whether the rank r is, or is not, 0. The problem of determining integers n, which are the area of a right-angled triangle having rational sides, goes back to Arabic manuscripts of more than a thousand years ago. This problem is the basis of the *Liber Quadratorum* (1225) of Leonardo of Pisa (*Fibonacci*). We require Pythagorean triples (a, b, c) so that $\frac{1}{2}ab = n$, or — and this is the same thing after setting $x = c^2/4$, $x \pm n = (a \pm b)^2/4$ — a rational square x so that both $x + n$ and $x - n$ are such squares; that is, n is the common difference of a three-term arithmetic progression of squares.† Traditionally, the successful n are called *congruent* numbers. For example (Fibonacci), 5 is congruent by virtue of the Pythagorean triple $(\frac{3}{2}, \frac{20}{3}, \frac{41}{6})$, or, equivalently, because of the square $x = 11\frac{97}{144} = (\frac{41}{12})^2$. Much more painfully, $n = 157$ is actually congruent because of an x with denominator of some 100 digits!

The elliptic curve reporting that all of $x - n$, x, and $x + n$ are squares, of course is $y^2 = x^3 - n^2x$. Here it is known that when n is prime and $n \equiv 3 \pmod 8$, then the rank is 0; the rank is at most 1 if n is prime and $n \equiv 5$ or $7 \pmod 8$; it is at most 2 if n is prime and $n \equiv 1 \pmod 8$. Finding the solution $(-4, 6)$ for $y^2 = x^3 - 5^2x$ proves that its rank is at least 1, and therefore 1. But Fermat could not have decided whether the rank is 0 or 1 in the case $n = 157$ since the smallest solution lies well outside any region in which he — or we — could hope to search for solutions. To decide in general even whether the rank is nonzero, one needs some new principle.

The idea turns out to be to consider the equation modulo p; that is, for each prime p to ask for the number N_p of pairs $(x, y) \pmod p$ with $y^2 \equiv x^3 + ax + b$. On crude statistical grounds — for each pair the probability of the congruence holding is $1/p$ — one guesses that simply

†Fermat could show by descent that one cannot have four squares in AP; thus the example 1, 25, 49 is an extreme case. The beauty of the group structure of the rational points on elliptic curves is that it both allows one to find all solutions *and* to show that there are no nontrivial solutions (if there aren't any).

$N_p \sim p$. Indeed, in 1933 Hasse established that $|N_p - p| < 2\sqrt{p}$ — known as the "Riemann Hypothesis for elliptic curves".

On collecting all this *local* data one hopes to obtain *global* information, that is, information on the rational solutions. I will talk about "local and global" in a later lecture, but let me just say here that the interrelationships turn out to be extremely striking, both actually and conjecturally.

But what does this all mean for Fermat's Last Theorem? We have learned that Faltings' proof of the Mordell Conjecture confirms that each equation $x^n + y^n = z^n$ with $n > 4$ has at most finitely many solutions, and that those solutions — if any, are probably just sporadic accidents. That does make it a little remarkable that there never is a nontrivial solution for $n > 2$. So it might make sense to look for additional structure. Indeed, let's contemplate the very simple equation $A + B = C$ in integers A, B, and C and then the elliptic curve $y^2 = x(x - A)(x + B)$, observing that its discriminant is essentially $(ABC)^2$. For a while there has been reason to guess that the discriminant should not be more than a small power (6 or so) of the product $\prod_{p|ABC} p$ of the different primes dividing A, B, and C. The truth of this conjecture immediately entails Fermat's Last Theorem, because

$$\prod_{p|x^n y^n z^n} p \le |xyz|,$$

as well as conquering a host of other inaccessible problems, such as that an equation $x^u - y^v = k$, with k composed from a nominated set of primes, has only small solutions in the integer exponents u, v and in relatively prime integers x and y.

The ABC-conjecture is still just that, a conjecture. However, Ribet was able to show in 1987, by proving a conjecture of Serre, that in the very special case $A = x^p$, $B = y^p$, $C = z^p$ the truth of the FLT followed from yet more fundamental remarks of Taniyama, Weil, and Shimura, and those for just the special case of *semistable* elliptic curves. At that time I heard Tate lecture on these matters at an AMS meeting at Arcata. It was exciting to see that our general views on elliptic curves entailed the FLT, but insofar as the various conjectures did not appear to be particularly accessible, we seemed little nearer to actually dealing with Fermat's Last Theorem.

I've been chatting with Don Zagier, the Australian Mathematical Society's 1993 Mahler Lecturer, and a visitor to ceNTRe. Other than for the concluding thoughts, virtually all this material is my vulgarization of ideas taken from his remarks and from his articles 'Lösungen von Gleichungen in ganzen Zahlen' [*Miscellanea Math.* (1991), pp. 311-326] and 'Elliptische Kurven: Fortschritt und Anwendungen' [*Jber. Deutsch. Math.-Verein* **92** (1990), pp. 58-76].

Notes and Remarks

VI.1 Diophantus was way ahead of his time. Also, upon his rediscovery in Europe, he was rather ahead of the then mathematics. He copes with the number 0, has no fear of negative numbers, and computes freely with rationals. Moreover, he even employs a notation that at least incorporates the now traditional unknown "x"; but he is restricted to just one unknown. Don Zagier explains that Diophantus writes $2x^3 - 3x^2 = 4$ as

$$K^{\Upsilon}\bar{\beta} \pitchfork \Delta^{\Upsilon}\bar{\gamma}\iota\overset{o}{M}\bar{\delta}\,.$$

Here $\bar{\beta}$, $\bar{\gamma}$, and $\bar{\delta}$ of course denote the numbers 2 , 3 and 4; K^{Υ} (kubos), Δ^{Υ} (dynamos), and $\overset{o}{M}$ (monad) represent x^3, x^2, and $x^0 = 1$ respectively. This is not quite as good as our notation but not bad even for the later times in which the experts might have had to learn factorizations such as

$$\mathrm{XCI} = \mathrm{VII} \times \mathrm{XIII}\,.$$

VI.2 We're not going to get a lot of pleasure from trying the chord and tangent method on the curve $x^3 + y^3 = 1$, because we know it has no nontrivial rational points. On the other hand, there is the celebrated story of Ramanujan and the taxicab, immortalized by Hardy: "I remember once going to see him when he was lying ill at Putney. I had ridden in taxicab n° 1729 and remarked that the number seemed to me rather a dull one, and that I hoped it was not an unfavorable omen. 'No,' he replied, 'it is a very interesting number; it is the smallest number expressible as a sum of two cubes in two different ways.'" [S. Ramanujan [*Collected Papers*, (New York: Chelsea, 1962), p. xxxv] The curve $x^3 + y^3 = 1729$ has the *integral* points $(1, 12)$ and $(10, 9)$. Incidentally, apropos of Ramanujan I quite commend Robert Kanigel, *The Man who Knew Infinity* (New York: Scribner's/Macmillan, 1991).

In any case, we can readily use the two solutions to $x^3 + y^3 = 1729$ to find yet further ones. The line through the cited solutions is $3y = -x + 37$, and using it to eliminate y in the equation we get $27x^3 + (-x + 37)^3 = 27 \cdot 1729$, which is $26x^3 + 111x^2 - 4107x + 3970 = 0$.

Thus the sum of the x coordinates of the three places this line cuts the curve is $-111/26$, and since we know two of those coordinates it follows that the third is $-397/26$. Substituting this back into the line, we find the corresponding y coordinate and learn that $(-397)^3 + 453^3 = 26^3 \cdot 1729$. Now there's no point in getting overexcited and trying this trick once more with the points $(-\frac{397}{26}, \frac{453}{26})$ and $(1, 12)$ because all we do, of course, is to find $(10, 9)$ again. But not to worry. Switching the coordinates of course yields a point $(\frac{453}{26}, -\frac{397}{26})$ on the curve, and using *it* with $(1, 12)$ provides a new solution with the yet nastier coordinates $(\frac{2472830}{187953}, -\frac{1538423}{187953})$. And so on. In this example, addition of points P and Q on the curve is effected by determining the third point at which the line $\overline{PQ}$ cuts the curve *and* exchanging coordinates. The exchange is a matter of replacing a point R such that $P + Q + R = 0$ by $-R = P + Q$. One sees this by noting that the third intersection of a line through (a, b) and (b, a) is at infinity. That is, on writing $x^3 + y^3 = 1729z^3$, one finds that the third point is $(1, -1, 0)$.

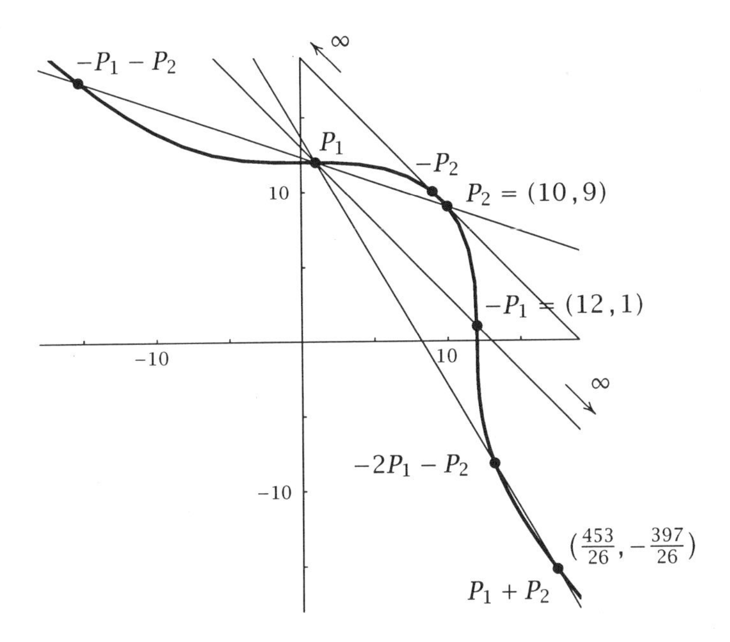

Rational points on the elliptic curve: $x^3 + y^3 = 1729$

By using the tangent at a point P one similarly computes the point $2P$. One might experiment with an even simpler case, say the point $P(2, -1)$ on $x^3 + y^3 = 7$. In this case, illustrated on the next page, $3P = (-\frac{17}{38}, \frac{73}{38})$.

I have fiercely borrowed this example from material 'Oefeningen rond Fermat', by Frits Beukers, used at a national Fermat Day (Fermatdag) at Utrecht on 6 November, 1993.

VI.3 In a similar spirit one applies the chord and tangent method to elliptic curves $y^2 = f(x)$, with $f(x)$ a cubic in x, by finding a third point on the line through P and Q and then *changing the sign* of the y coordinate of the third point.

VI.4 One of my readers, René Hutchins, helpfully reminded me of the congenial example $y^2 = x^3 - 43x + 166$, which contains the point $P(3, 8)$. It's now easy to successively find the points $2P(-5, -16)$, $4P(11, 32)$, and $8P(3, 8)$; thus noticing that P is a torsion point of order 7 on the curve.

VI.5 The case $y^2 = f(x)$ with f a polynomial of degree 4 is rather more interesting. I'll show in Lecture IX that there always is a transformation bringing us back into the cubic case; but let me here, guided by Fermat's

Inventum Novum and remarks of Beukers *op. cit.*, mention the example $y^2 = x^4 + a$ relevant to the case $n = 4$ of Fermat's Last Theorem. Given points P and Q on the curve, Fermat's trick is to construct a parabola $y = x^2 + bx + c$ through the two points. On substituting for y, Fermat found, as do we, that the term in x^4 cancels. So Fermat, as above did we, can find the rational x coordinate of the third point of intersection. All these phenomena must have been a miracle to Fermat. After all, he could hardly have appreciated that the term in x^4 disappears because the parabola and the given curve have intersection of high multiplicity at ∞. I understand the principles to some extent and will try to explain them in future lectures, yet I too cannot quite shake my sense of wonder. And my admiration for Fermat is unbounded — even though he made too hasty a marginal note.

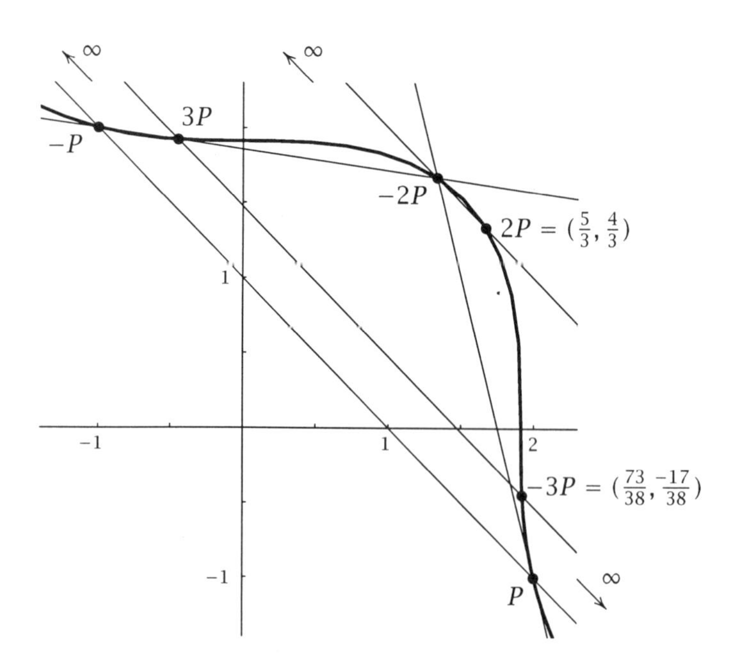

Rational points on the elliptic curve: $x^3 + y^3 = 7$

Copyleft Kristoffer Rose and Ross Moore, 1995

VI.6 I should emphasize that the notion of *genus* has little to do with degree, mainly because there are many equations of quite different degree that define—or are *models* for—the same curve. In different language, there are quite drastic transformations of an equation that among other things do not disturb the genus of the curve being modeled. But loosely speaking, my degree remark stands. Still, the remark is horribly loose. Thus the Fermat curve $x^4 + y^4 = z^4$ is of genus 3; the curve $x^4 + y^2 = z^4$, cleverly considered by Fermat, leads to a curve of genus 1. In general, the

Fermat curve $x^n + y^n = z^n$ has genus $\frac{1}{2}(n-1)(n-2)$. I talk about these matters in Lecture IX.

VI.7 The term "congruent", as in "congruent number" n, arises from the fact that such numbers are precisely the common difference (*congruum*) in a three-term arithmetic progression of rational squares. I am indebted to Noam Elkies for reminding me that early history of the congruent number problem appears in the Preface and Chapter XVI of L. E. Dickson, *History of the Theory of Numbers*, Vol. II: *Diophantine Analysis* (New York: Chelsea, 1971). Similarly, Neal Koblitz [*Introduction to Elliptic Curves and Modular Functions* (New York: Springer-Verlag, 1984)] uses the congruent number problem as his central motivation, much as here I am using Fermat's Last Theorem. But notwithstanding being readable, Neal's book is a reliable and serious mathematics text.

VI.8 Frank Calegari points out to me that to show that there are no four rational squares in AP, as Fermat did, amounts to proving that the group of rational points on the elliptic curve $y^2 = x^4 + 8x^3 + 2x^2 - 8x + 1$ has rank zero. In practice, the easiest way to prove anything is to look it up. Here it suffices to notice that the given curve is birationally equivalent to $y^2 = x^3 - x^2 - 4x + 4$. The curve is of conductor $N = 24$ and then John Cremona's tables, or those in Antwerp IV, readily verify that the rank is indeed zero. Mind you, Fermat's proof by descent did not just rely on table lookup.

VI.9 Zagier ['Oplossingen van vergelijkingen in rationale getallen', once again at the Utrecht *Fermatdag*] makes the following beautiful remark concerning the rank of the group of rational points on an elliptic curve. Given the conjecture of Taniyama, Weil, and Shimura, one may associate a certain integer S with each elliptic curve. According to the celebrated conjecture of Birch and Swinnerton-Dyer, the curve has infinitely many rational points exactly when $S = 0$. The conjecture, established in a wide variety of cases, is useful because S can be calculated in bounded time. For example, in the case of the elliptic curves $x^3 + y^3 = m$ one obtains the values

$$\begin{array}{lcccccccccccc} m: & 1 & 2 & 3 & 4 & 5 & 6 & 7 & \cdots & 381 & 382 & 383 & \cdots \\ S: & 1 & 1 & 1 & 1 & 1 & 0 & 0 & \cdots & 4 & 0 & 4 & \cdots \end{array} .$$

It follows that $7 = 2^3 + (-1)^3$ is not surprising in that 7 must be the sum of two cubes; as must 6, though $6 = \left(\frac{17}{21}\right)^3 + \left(\frac{37}{21}\right)^3$ is a little harder to see. In the case $m = 382$ it really does turn out to be easier to compute $S = 0$ than to solve the equation $x^3 + y^3 = 382$. The smallest solution has x and y with denominator

$$8122054393485793893167719500929060093151854013194574;$$

the numerators are even larger. In any case, the Birch–Swinnerton-Dyer Conjecture is, subject to its full validity, the final link in providing a complete answer to the question as to which polynomial equations in two variables defined over $\mathbb{Q}$ — which curves defined over rationals — have infinitely many rational solutions. For rational curves this is always the case if there is any solution; for curves of general type this is never so by virtue of Faltings' proof of the Mordell Conjecture; and for elliptic curves it is so when and only when the number S is equal to 0.

One can compute S by (actually some additional small rational factor, the product of the Tamagawa numbers, should be inserted here, but that doesn't affect the interesting case when S vanishes) Tamagawa numbers

$$S = \frac{1}{\Omega} \cdot \frac{2}{N_2+1} \cdot \frac{3}{N_3+1} \cdot \frac{5}{N_5+1} \cdot \cdots \cdot \frac{p}{N_p+1} \cdot \cdots .$$

Such an asymptotic formula does allow the computation of S given that it is an integer—in fact a square. The product is over primes $p = 2, 3, 5, \ldots$ and N_p is, as above, the number of pairs (x, y) of integers $0 \le x, y < p$ so that $p \mid f(x, y)$. Finally, Ω is an "elliptic integral", usefully instanced by

$$\Omega = \int_{-1}^{\infty} \frac{dx}{\sqrt{x^3+1}}$$

for the curve $y^2 = x^3 + 1$. However, all this presumes that the elliptic curve is a Weil curve; that is, it can be parametrized by modular functions. Otherwise, one has no business claiming anything about S at all.

VI.10 Did you notice the clever hyphenation in "Birch–Swinnerton-Dyer"? TeX makes that easy and minimizes the surprise that innocents once would feel on meeting Professor Sir Peter Swinnerton-Dyer and realizing that here was two-thirds of a celebrated conjecture in just the one man.

VI.11 But let me explain that mysterious, yet very important, remark that the idea turns out to be to consider the equation modulo p; that is, for each prime p to ask for the number N_p of pairs $(x, y) \pmod{p}$ with $y^2 \equiv x^3 + ax + b$. Suppose that we were to do this sort of thing for the rather more harmless curve $x^2 + y^2 = 1$, recalling that the solutions are parametrized by $((1-t^2)/(1+t^2), 2t/(1+t^2))$. Of course $N_2 = 2$. Modulo p there's a difference according as -1 is a square modulo p or not. In the first case, there are just $N_p = p - 1$ possibilities for $t \pmod{p}$, yielding precisely that many different solutions. In the other case, asymmetry yields $N_p = p + 1$ solutions. By quadratic reciprocity, respectively $p \equiv 1$ and $p \equiv -1 \pmod 4$ provide the two cases. Finally, the length $N_{\mathbb{R}}$ of the real solutions is 2π. That yields the product

$$L = \prod \frac{p}{N_p} = \prod_{p\equiv 1} \frac{1}{1-\frac{1}{p}} \prod_{p\equiv -1} \frac{1}{1+\frac{1}{p}} = 1 - \frac{1}{3} + \frac{1}{5} - \frac{1}{7} + \frac{1}{9} - \cdots = \frac{\pi}{4},$$

and, noticing that the number of integer solutions is $N_{\mathbb{Z}} = 4$, we have

$$\frac{N_p}{p} \cdot N_{\mathbb{R}} = 2N_{\mathbb{Z}} .$$

Similarly, Pell's equation: $x^2 - 2y^2 = 1$ has $p-1$ solutions modulo p if p is ± 1, and $p+1$ solutions if it is $\pm 3 \pmod 8$. It follows that

$$\prod \frac{N_p}{p} = \frac{2\sqrt{2}}{\log(3+2\sqrt{2})}.$$

In this case $N_{\mathbb{R}}$ is infinite, but so is $N_{\mathbb{Z}}$. However there's a group law on solutions given by $(x,y)\cdot(x',y') = (xx'+2yy', xy'+x'y)$. The map $(x,y) \mapsto \log(|x+y\sqrt{2}|)$ makes this an addition and now comparing lengths of the real and integer solutions yields $N_{\mathbb{R}}/N_{\mathbb{Z}} = \log(3+2\sqrt{2})$. So

$$\prod \frac{N_p}{p} \cdot \frac{N_{\mathbb{R}}}{N_{\mathbb{Z}}} = \sqrt{8}.$$

Of course, 4, respectively 8, is just $|\Delta|$, where Δ is the discriminant of the respective polynomials X^2+1 and X^2-2.

The point is now, in Henri Darmon's fine wording, that one may expect a "resonance" between all this and the elliptic case. That appropriately interpreted resonance becomes the conjectures of Birch–Swinnerton-Dyer.

VI.12 One speaks of the "Riemann Hypothesis" for curves, because the assertion is that certain quantities have absolute value $\sqrt{p} = p^{1/2}$, the operative thing being that exponent. More vividly, were one to replace the variable by p^s, then at the "roots" one has $s = \frac{1}{2} + it$ in perfect analogy with the classical Riemann Hypothesis, of which more later.

VI.13 Zagier's example, as given above, is an interesting illustration of the power of the conjectures that dominate the subject. Given truth of the conjectures, all sorts of computations are possible. Their success provides compelling evidence in support of the conjectures. If we were talking physics, these "theories" would be deemed to be quite well established. But we're talking about the ideal world, not just the real one. As mathematicians, all we can do is to pat our stomachs, or whatever else we deem to be the site of our beliefs, and reconfirm our "feelings". It is in this context that Wiles' result is so important. One can never be certain about one's stomach; it may only be wind. Wiles confirms that the views that have driven our investigations are *true* and are not just some astonishingly effective approximation that happens to work for the relatively small numbers accessible to us and our computers.

VI.14 As we'll hear again and again later, the conjecture of Taniyama, Weil, and Shimura asserts that elliptic curves may be parametrized not just by elliptic functions — this is what makes them elliptic in the first place — but also by modular functions. Eventually we'll learn what defines such beasts, but for the moment let me quote an example, once again from Zagier. The Fermat curve $x^4 + y^4 = 1$ has a parametrization by the functions

$$x(t) = \frac{1 - 2t^4 + 2t^{16} - 2t^{36} + \cdots}{1 + 2t^4 + 2t^{16} + 2t^{36} + \cdots},$$

$$y(t) = \frac{2t + 2t^9 + 2t^{25} + 2t^{49} + \cdots}{1 + 2t^4 + 2t^{16} + 2t^{36} + \cdots},$$

in which the exponents are the even and odd squares. These functions also have delightful representations as products

$$x(t) = \left(\frac{1 - t^4}{1 + t^4} \cdot \frac{1 - t^{12}}{1 + t^{12}} \cdot \frac{1 - t^{20}}{1 + t^{20}} \cdot \cdots\right)^2$$

$$y(t) = 2t\left(\frac{1 + t^8}{1 + t^4} \cdot \frac{1 + t^{16}}{1 + t^{12}} \cdot \frac{1 + t^{24}}{1 + t^{20}} \cdot \cdots\right)^2$$

wherein the exponents are now the odd and even multiples of 4. Modularity is entailed by the identities $x(t) = x(u)$ and $y(t) = y(u)$ valid whenever $(\log u)(\log t) = -\pi^2/16$.

VI.15 I alluded to the equation $x^u - y^v = k$. It is a generalization of Catalan's equation $x^u - y^v = 1$, concerning which Catalan conjectured in 1844 that its only solution in integers x, y, u, v all greater than 1 is given by $3^2 - 2^3 = 1$. The equation is the subject of a recent book, *Catalan's Conjecture*, by Paulo Ribenboim (New York: Academic Press, 1994). In reviewing that volume, [*Bull. Amer. Math. Soc.* **32** (1995)], Alan Baker makes the comment that "the conjecture gained considerable notoriety, and it became plain that it represented a challenge to number theorists somewhat akin to Fermat's Last Theorem."

Since the problem seemed rather less accessible than the FLT it was a real surprise when, in 1974, Tijdeman, applying a newly refined version of Baker's inequalities for linear forms in logarithms, found an effectively computable bound for the four variables. That doesn't quite settle the matter. Current refinements of the inequalities still yield a bound of 10^{13} or so for $\min(u, v)$; conversely, algebraic methods based on theorems of Inkeri yielding congruences of the genre $u^{v-1} \equiv 1 \pmod{v^2}$ now give a lower bound of a 100 or so for u, or v, occurring in any nontrivial solution.

With that all said, it may seem curious that nothing much at all has been proved for $x^u - y^v = k$ if $k \neq \pm 1$. Once again, one confidently conjectures that given k, or even just the prime factors of k, there are only small solutions; one usually excludes the fairly trivial case $u = v = 2$. I spent a great deal of time on that question back in the seventies, with nothing more to show for it than a p-adic generalization of Tijdeman's theorem giving a bound for the variables in $x^u - y^v = k^{uv}$ in terms of the prime factors of k.

VI.16 I should instantly confess to confusing matters somewhat by talking about an ABC-conjecture in the way that I do. It is indeed true that there is a theorem of Szpiro bounding the discriminant *vis-à-vis* the conductor, for elliptic curves defined over function fields. Its numerical analogue, known as Szpiro's Conjecture, would say the kind of thing I mention.

I'll explain later in Lecture XIV that there in any case is a remark — properly called the ABC-conjecture — of Masser and Oesterlé, once again based on an analogy with a theorem in function fields, whose truth would readily entail the FLT. That remark, however, has nothing to do with elliptic curves *per se*. Mind you, the ABC-conjecture proper approached directly seems quite inaccessible, whereas now, given Wiles' work, one somehow does not feel quite as confident of the inaccessibility of Szpiro's Conjecture. Mind you, that's silly given the interconnections of these various conjectures. Ribet's work really does depend on quite different principles. He *proves* that Frey curves cannot be modular thereby confirming that the FLT must surely hold, lest the Taniyama-Weil-Shimura Conjecture be false.

One can readily become confused as between theorem and conjecture in this subject. The trouble is that so many of the things spoken about elliptic curves seem to be conjectures, themselves contingent on yet other conjectures. Thus, for example, the Birch-Swinnerton-Dyer Conjectures speak about the value of a function at a point at which it is not even known to exist. Fortunately the Taniyama-Weil-Shimura Conjecture at least entails the existence of the L-function at that critical point.

John Tate at the Boston meeting, August 1995

Photograph by C. J. Mozzochi, Princeton

DIOPHANTI ALEXANDRINI ARITHMETICORVM LIBRI SEX,

ET DE NVMERIS MVLTANGVLIS LIBER VNVS.

CVM COMMENTARIIS C. G. BACHETI V. C. & obseruationibus D. P. de FERMAT *Senatoris Tolosani.*

Accessit Doctrinæ Analyticæ inuentum nouum, collectum ex varijs eiusdem D. de FERMAT Epistolis.

TOLOSÆ,
Excudebat BERNARDVS BOSC, è Regione Collegij Societatis Iesu.

M. DC. LXX.

The 1670 Reprint

title-page from *Diophanti Alexandrini Arithmeticorum libri sex* . . . , Tolosæ 1670.

Notes on Fermat's Last Theorem

LECTURE VII

ellipsis *the omission from a sentence of a word or words which would complete or clarify the construction.*

Today's story is a child's introduction to elliptic functions. Since I'll be covering several terms' coursework in a few pages — my remarks will be elliptical — we had best fasten our seatbelts.

When there is talk of periodic functions one thinks of $\sin 2\pi z$ and $\cos 2\pi z$; I've put in the 2π as a normalization so that these functions have *primitive* period 1, rather than some random amount ω, or whatever. The adjective "primitive" is there to acknowledge that they actually have periods $0, \pm 1, \pm 2, \ldots$, but fundamentally the period is 1 as said. Of course, we say that f has period ω if $f(z+\omega) = f(z)$ for all z. It is not a very sophisticated remark to observe that the *circular functions* are periodic because $e^{2i\pi z}$ is periodic; after all, they are just its imaginary and real parts, respectively.

To add that, in fact, *every* periodic function is periodic because it is itself a function of $e^{2i\pi z}$ is not quite trivial. In practice this manifests itself by reasonable periodic functions $g(z)$ having a Fourier expansion

$$\sum_{n=-\infty}^{\infty} c_n e^{2ni\pi z} = c_0 + \sum_{n=1}^{\infty} \left((c_n + c_{-n})\cos 2n\pi z + i(c_n - c_{-n})\sin 2n\pi z\right).$$

Mind you, it is worthwhile to pause to ask just how we might have known in the first place that $e^{2i\pi z}$ is periodic. Surely this is not obvious from its power series definition. Let me suggest two different "explanations". In the first we treat sin as original by computing the length of an arc of the circle $x^2 + y^2 = 1$ between the ordinates $y = 0$ and $y = \sin z$, and define sin by

$$z = \int_0^{\sin z} \frac{dy}{\sqrt{1-y^2}}.$$

Then the circularity of the circle entails the periodicity of sin.

I prefer the following illustration. Here we admit that $\sin \pi z$ has simple zeros exactly at $0, \pm 1, \pm 2, \ldots$, and — rather wildly thinking of it as just a polynomial of infinite degree — we factorize it and write

$$\sin \pi z = \pi z \prod_{n=1}^{\infty} \left(1 - \frac{z^2}{n^2}\right).$$

Of course that multiplier π (which, after all, might have been any decent function that never vanishes) needs rather calmer justification.*

With this evil deed done, we acknowledge that we are frightened of products, so we take the logarithm; and being bothered by logarithms, we differentiate. That yields

$$\pi \cot \pi z = \frac{1}{z} - \sum_{n=1}^{\infty} \left(\frac{1}{n-z} - \frac{1}{n+z}\right).$$

Unfortunately, as we catch our breath, we see that this is a mildly nasty partial fraction expansion† in that it only converges conditionally — that is, on condition that we don't muck about with those parentheses. So we differentiate again and contemplate

$$\pi^2 \operatorname{cosec}^2 \pi z = \sum_{n=-\infty}^{\infty} \frac{1}{(n-z)^2}$$

and see that it shouts its periodicity. If we now backtrack carefully, we are done.

Of course I've told this story to motivate its generalization. I'd better also announce a principle. Loosely speaking, a function is "good" if it is a convergent sum of good functions. That's why the sums above are good if we stay away from their *poles* — the points at which a term blows up, the integers $\mathbb{Z}$ in our examples. All this is given that "good" — *analytic* — sort of means "differentiable"; or better said, that the function may be expanded as a convergent power series. At the poles our functions — examples of *meromorphic* functions — are quite bad but not very bad.

*The trick is to notice that De Moivre's theorem, and then writing $1-\sin^2\theta$ for $\cos^2\theta$, allows us to write

$$\sin(2m+1)\theta = \sin\theta P_{2m}(\sin\theta)$$

with P_{2m} a polynomial of degree $2m$ and constant term $(2m+1)$. Of course, its $2m$ zeros are $\pm\sin\big(\pi k/(2m+1)\big)$ for $k = 1, 2, \ldots, m$, so we have

$$\sin(2m+1)\theta = (2m+1)\sin\theta \prod_{k=1}^{m}\left(1 - \sin^2\theta \Big/ \sin^2\frac{\pi k}{2m+1}\right).$$

Now there is little more to do than to set $\theta = \pi z/(2m+1)$ and to let m go to ∞.

†But now I can tell about Euler's evaluation of $\zeta(2k)$. We obtain

$$i\pi z \cot i\pi z = \pi z + \frac{2\pi z}{e^{2\pi z}-1} = \pi z + \sum_{m=0}^{\infty} \frac{B_m}{m!}(2\pi z)^m$$

$$= 1 + 2\sum_{k=0}^{\infty}(-1)^k \sum_{n=1}^{\infty} \frac{1}{n^{2(k+1)}} z^{2(k+1)}.$$

Comparing coefficients of z^{2k}, we have the claim ending Lecture III.

They just take the value ∞; in other words, their reciprocal is 0. That's not too bad. Incidentally, after taking a logarithm, infinite products are just infinite sums. I'll try only to deal with functions that are good everywhere, except possibly for the odd pole; such pretty good functions are called *meromorphic* functions. These functions may well be very bad at $z = \infty$; that's permitted. A function without any singularity in the finite complex plane is called an *entire* function. I will use the fact that entire functions, excepting of course the constants, are not bounded.

If a nonconstant good function f has essentially different periods 1 and τ, then τ must be a dinkum* complex number. Certainly, τ cannot be rational, either because then 1 is not a primitive period or, anyhow, because then the periods are not essentially different. Nor can τ be real irrational, because then there are $\mathbb{Z}$-linear combinations of 1 and τ that are arbitrarily small, and f would have to be just a miserable constant. Thus τ must be $\mathbb{R}$-linearly independent of 1 and we may choose it in the upper half-plane $\mathcal{H}$, that is, with positive imaginary part.

Just as $\mathbb{Z}$ is all the periods of $e^{2i\pi z}$, so all the periods of a doubly periodic function with primitive periods 1 and τ is the *lattice*

$$\Lambda = \{\omega = n\tau + m : n, m \in \mathbb{Z}\}.$$

Then the sum

$$\wp'(z) = 2 \sum_{\omega \in \Lambda} \frac{1}{(\omega - z)^3}$$

defines a meromorphic doubly periodic function with period lattice Λ. One confirms that just as $\sum n^{-k}$ converges provided that $k > 1$, so $\sum' \omega^{-k}$ converges absolutely if $k > 2$. The $'$ tells one to omit the term with $\omega = 0$. Integrating, we obtain

$$\wp(z) = \frac{1}{z^2} + \sum_{\omega \in \Lambda}{}' \left(\frac{1}{(\omega - z)^2} - \frac{1}{\omega^2} \right).$$

The periodicity of the Weierstrass $\wp$-function follows by observing that certainly $\wp(z + \omega) - \wp(z)$ is constant. Now let ω be a primitive period and take $z = -\frac{1}{2}\omega$ to see that the constant is zero because $\wp(z)$ is an even function of z.

Just as we produced the Riemann ζ-function in expanding $\operatorname{cosec}^2 \pi z$ as a power series, so expanding $\wp'(z)$ yields the Eisenstein series G_{2k},

$$\wp'(z) = -2z^{-3} + 6G_4 z + 20G_6 z^3 + \cdots \quad \text{where} \quad G_{2k} = G_{2k}(\tau) = \sum_{\omega \in \Lambda}{}' \frac{1}{\omega^{2k}}.$$

There is now nothing for it other than brute computation to discover that

$$(\wp'(z))^2 = 4(\wp(z))^3 - 60G_4\wp(z) - 140G_6$$

---

* *Macquarie Dictionary*: **dinkum** true, honest, genuine: as in *dinkum Aussie.*

because the difference of the sinister and dexter sides of this equation is a doubly periodic function without poles, hence bounded, and furthermore, it vanishes at $z = 0$.

An amusing corollary is that the G_{2k} must all be polynomials in just G_4 and G_6. Thus, for example, we must have

$$\left(\sum_{\omega\in\Lambda}{}' \frac{1}{\omega^4}\right)\left(\sum_{\omega\in\Lambda}{}' \frac{1}{\omega^6}\right) = c\sum_{\omega\in\Lambda}{}' \frac{1}{\omega^{10}},$$

a child's dream. Here c is an absolute constant, independent of Λ; it is easy to determine but I have been too lazy to compute it.

Put $y = \wp'(z)$ and $x = \wp(z)$. Then

$$y^2 = 4x^3 - g_2x - g_3 \quad \text{with} \quad g_2 = 60G_4 \text{ and } g_3 = 140G_6$$

is the equation of an *elliptic curve*. In truth, this equation has nothing much to do with ellipses. The background is that the integral that gives the length of an arc of an ellipse gives rise to a doubly periodic function; thence the terminology whereby the Weierstrass $\wp$-function (and indeed all doubly periodic meromorphic functions) are called *elliptic functions*.

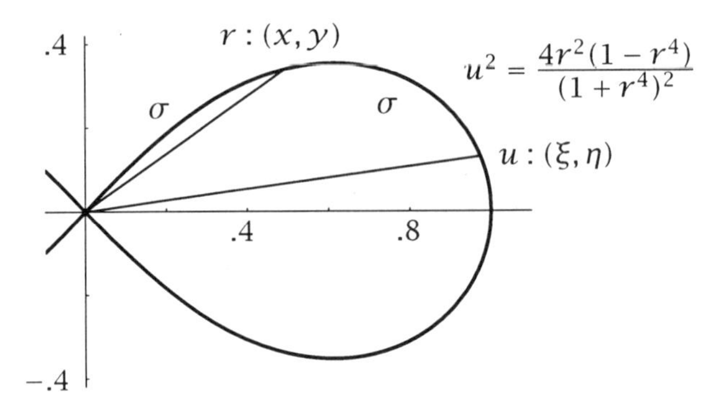

Doubling of arc length on the lemniscate

All this goes back to Fagnano's *rectification* of the *lemniscate*, the locus of a point which moves such that the product of its distances from two given points remains a positive constant, which leads to the integral

$$\int_w^\infty \frac{dr}{\sqrt{1-r^4}}.$$

Our curve is an *elliptic* curve because it may be *parametrized* by elliptic functions. I will eventually try to explain how it became believed that these curves can also be parametrized by modular functions, and how that entails Fermat's Last Theorem.

My earlier remarks readily provide the factorization

$$\begin{aligned}4x^3 - g_2 x - g_3 &= 4\left(x - \wp(\tfrac{\tau}{2})\right)\left(x - \wp(\tfrac{1}{2})\right)\left(x - \wp(\tfrac{\tau+1}{2})\right)\\ &= 4(x-e_1)(x-e_2)(x-e_3)\,,\end{aligned}$$

showing that the cubic has distinct zeros. Thus its discriminant $g_2^3 - 27g_3^2$, the square of the difference product $(e_1-e_2)(e_2-e_3)(e_3-e_1)$, is non zero.

Now a word about the lattice Λ. Let a, b, c and d be integers so that $ad - bc = 1$. Then the ordered pair $(a\tau + b, c\tau + d)$ generates the same lattice as did $(\tau, 1)$ and $(a\tau + b)/(c\tau + d)$ is still in $\mathcal{H}$, the upper half-plane. Set

$$M = \begin{pmatrix} a & b \\ c & d \end{pmatrix} \qquad \text{and define} \qquad M\tau = \frac{(a\tau + b)}{(c\tau + d)}\,.$$

Then one readily verifies that

$$G_{2k}(M\tau) = (c\tau + d)^{2k} G_{2k}(\tau)\,.$$

The Eisenstein series are the basic examples of *modular forms*, that is, of functions exhibiting an invariance under transformations by a discrete group. In this case the group is the full modular group $\Gamma = \mathrm{SL}_2(\mathbb{Z})/\{I, -I\}$ of linear fractional transformations of determinant 1.

Elliptic functions parametrize elliptic curves. Following a suggestion of Taniyama in the early fifties and subsequent work of Shimura and then of Weil, it became believed by the late sixties that such curves, if defined over the rationals $\mathbb{Q}$, also must be parametrized by certain modular functions.

Yet in the mid-eighties Gerhard Frey remarked that if $a^p + b^p = c^p$ and $p \geq 5$ say, then the elliptic curve $y^2 = x(x - a^p)(x + b^p)$ — note that its discriminant is essentially $(abc)^{2p}$, so the definition is symmetric — could *not* be modular. That was proved by Ken Ribet in 1986, showing that the truth of the Modularity Conjecture entails Fermat's Last Theorem.

André Weil, *Elliptic Functions According to Eisenstein and Kronecker*, Ergebnisse der Mathematik **88** (New York: Springer-Verlag, 1976) makes instructive reading.

Notes and Remarks

VII.1 The principle that one should manifest symmetries of a function by expanding it accordingly, hence Fourier expansions, is of the greatest importance. We will use it again and again. Let me make just two remarks here. The first is that it is easy to compute the Fourier coefficients c_n, by

$$c_n = \int_0^1 g(z)e^{-2ni\pi z}\,dz\,.$$

The second has us first write $q = e^{2i\pi z}$ so that the expansion gets the appearance $\sum c_n q^n$ of a power series in q. We see that if there are lots of $c_n \neq 0$ for $n < 0$, then nasty things happen at $q = 0$. But should it happen to happen — as it will for many of the functions that will interest us in the sequel — that $c_n = 0$ for all $n < 0$, then our function is regular at $q = 0$. This means that we can say that our function is regular at infinity, more precisely, at $z = i\infty$. Thus in particular, the Fourier expansion provides a clear glimpse of what is happening at infinity.

VII.2 Notwithstanding that fine line about circularity of the circle and periodicity of sin, I really have to prefer the second "explanation". For what can I have meant by the first? The trouble is that I am trying to talk as if you know next to nothing about integration in the complex plane, whereas this remark really requires some minimal understanding that the path from 0 to $\sin z$ may be long and circuitous. I think I'll let it pass. *Mea culpa*, I sinned.

VII.3 I shouldn't just mention Fagnano's integral without a few details. A lemniscate is the locus of a point moving so that the product of its distances from two fixed points, say $(\pm a, 0)$ is a constant, say c^2. Writing $r^2 = x^2 + y^2$ as usual, the square c^4 of the product of the distances of (x, y) from the given points is $(r^2 + a^2)^2 - 4a^2x^2$. Now taking the special case $c = a$, then normalizing so that $2a^2 = 1$, we obtain parametric equations $2x^2 = r^2 + r^4$ and $2y^2 = r^2 - r^4$. It is now easy to conclude that $ds = (1 - r^4)^{\frac{-1}{2}} dr$ as asserted. This matter is the opening topic of Siegel's *Topics in Complex Function Theory* (New York: Wiley–Interscience, 1969). Siegel explains how one stumbles onto a duplication formula and thence an addition formula on the lemniscate. It is just such an addition formula that accounts for the group law on elliptic curves. It's perhaps amusing to note that Euler's work on the lemniscatic function in 1752 was quickly followed by his settling the case $n = 3$ of Fermat's Last Theorem, in 1753.

VII.4 More exactly, the square of the difference product of the zeros of the cubic $4x^3 - g_2x - g_3$ is $\frac{1}{16}(g_2^3 - 27g_3^2)$. Let's see how to see such a thing by considering a cubic $(x - \alpha)(x - \beta)(x - \gamma)$. Then the square of the difference product of the roots $(\alpha - \beta)(\beta - \gamma)(\gamma - \alpha)$ is given by the determinant product

$$\begin{vmatrix} 1 & 1 & 1 \\ \alpha & \beta & \gamma \\ \alpha^2 & \beta^2 & \gamma^2 \end{vmatrix} \cdot \begin{vmatrix} 1 & \alpha & \alpha^2 \\ 1 & \beta & \beta^2 \\ 1 & \gamma & \gamma^2 \end{vmatrix}$$

$$= \begin{vmatrix} 3 & \alpha + \beta + \gamma & \alpha^2 + \beta^2 + \gamma^2 \\ \alpha + \beta + \gamma & \alpha^2 + \beta^2 + \gamma^2 & \alpha^3 + \beta^3 + \gamma^3 \\ \alpha^2 + \beta^2 + \gamma^2 & \alpha^3 + \beta^3 + \gamma^3 & \alpha^4 + \beta^4 + \gamma^4 \end{vmatrix}.$$

It's traditional to set $s_m = \sum \alpha^m$ for the sum of the mth powers of the zeros, denoting the symmetric functions respectively by 1, σ_1, σ_2, ...

That is, in the case of the polynomial $4x^3 - g_2x - g_3$ we have that $\sigma_1 = 0$, $\sigma_2 = -\frac{1}{4}g_2$, and $\sigma_3 = \frac{1}{4}g_3$.

Obviously, $s_1 = \sigma_1 = 0$ and $s_2 = (s_1)^2 - 2\sigma_2 = \frac{1}{2}g_2$. But we know that $4\alpha^3 - g_2\alpha - g_3 = 0$ and similarly for the other zeros. So we plainly have $4s_3 = g_2s_1 + 3g_3$, which is $s_3 = \frac{3}{4}g_3$. Similarly, $4s_4 = g_2s_2 + g_3s_1$, so that $s_4 = \frac{1}{8}g_2^2$. It is now easy to evaluate the determinant.

VII.5 However, perhaps it's more correct to say that the discriminant of the polynomial $4x^3 - g_2x - g_3$ is $4^2(g_2^3 - 27g_3^2)$. This arises as follows. Given polynomials f and g of degree n and m, respectively, it is simple linear algebra to confirm that there exist polynomials A and B of degree $m - 1$ and $n - 1$, respectively so that $Af + Bg = R$ is constant. The point is that this is a matter of solving $m + n - 1$ homogeneous equations in the $m + n$ unknown coefficients of the polynomials A and B. Moreover, we can solve so that the coefficients of A and B are polynomials, with integer coefficients, in the coefficients of f and g. The constant R is known as the *resultant* $R(f,g)$ of f and g. But plainly, if f and g have a common zero, then the constant R must vanish. The upshot is that if $f(x) = a_0(x - \alpha_1)\cdots(x - \alpha_n)$ and $g(x) = b_0(x - \beta_1)\cdots(x - \beta_m)$, then we must have

$$R(f,g) = a_0^m b_0^n \prod_{i=1}^{n}\prod_{j=1}^{m}(\alpha_i - \beta_j) = a_0^m \prod_{i=1}^{n} g(\alpha_i).$$

If g is the derivative f' of f, we obtain the discriminant

$$R(f,f') = a_0^{2m-1}\prod_{i=1}^{n}\prod_{j\neq i}(\alpha_i - \alpha_j).$$

All this said, the definition of the discriminant of an elliptic curve

$$y^2 = 4x^3 - g_2x - g_3$$

is slightly different yet again; we'll come to it in a little while.

VII.6 I shouldn't leave these matters without mentioning the Newton–Raphson formulas relating the elementary symmetric functions and the sums of powers s_j. Very briefly, one defines the elementary symmetric functions σ_i by

$$\begin{aligned} f(x) &= (x - \alpha_1)(x - \alpha_2)\cdots(x - \alpha_n) \\ &= x^n - \sigma_1x^{n-1} + \cdots + (-1)^{n-1}\sigma_{n-1}x + (-1)^n\sigma_n. \end{aligned}$$

Then the formulas announce that for all $h = 0, 1, 2, \ldots$

$$\sigma_0s_h - \sigma_1s_{h-1} + \sigma_2s_{h-2} - \cdots + (-1)^{h-1}\sigma_{h-1}s_1 + (-1)^h h\sigma_h = 0.$$

Of course I am adopting a convention whereby $\sigma_k = 0$ if $k > n$.

As it happened I looked at B. L. van der Waerden's *Modern Algebra*, to be reminded of the proof of the formula, and was concerned when I found it to be just an exercise. However, the shape of the formula screams the following argument. One first notices that by logarithmic differentiation

$$x^{-1}\frac{f'(x^{-1})}{f(x^{-1})} = \frac{1}{1-\alpha_1 x} + \cdots + \frac{1}{1-\alpha_n x} = \sum_{m=0}^{\infty} s_m x^m .$$

Clearly, one now multiplies by $x^n f(x^{-1})$, obtaining $x^{n-1}f'(x^{-1})$, then just compares coefficients of x^h.

VII.7 I said quickly, without any attempt at explanation, that because the three zeros of $4x^3 - g_2x - g_3$ are given by the x coordinates of the points $\big(\wp(\frac{1}{2}\tau), 0\big)$, $\big(\wp(\frac{1}{2}), 0\big)$ and $\big(\wp(\frac{1}{2}(1+\tau)), 0\big)$, it follows that the three zeros are distinct. The fact to notice is that with $e = \wp(\frac{1}{2}\omega)$, ω a primitive period of $\wp(z)$ — thus in the present case the possibilities in the fundamental parallelogram are exactly $\omega = \tau$, or 1 or $1+\tau$ — the elliptic function $\wp(z) - e$ has a *double* zero at $\frac{1}{2}\omega$. That's clear because

$$(\wp'(z))^2 = (\wp(z) - e_1)(\wp(z) - e_2)(\wp(z) - e_3)$$

entails that its derivative $\wp'(z)$ also vanishes at $\frac{1}{2}\omega$. It's a basic fact that because $\wp(z) - e$ has just a double pole in its period parallelogram, then it has just two zeros in that parallelogram. Hence the e_i must be distinct. Incidentally, these three points $P_i = (e_i, 0)$, together with the point O at infinity, are precisely the "2-division points" on the curve. That is, they satisfy $2P_i = O$ in the sense described in Lecture VI.

VII.8 It's natural to talk about groups of invertible matrices (*group* means no more than the collection being closed under division). But the trouble is that we are actually interested in *linear fractional* transformations (what used to be known as Möbius transformations) $\tau \mapsto (a\tau + b)/(c\tau + d)$. Plainly, matrices M and $-M$ provide the same linear fractional transformation, hence my factoring out the subgroup $\{I, -I\}$ from the special linear group $\mathrm{SL}(2, \mathbb{Z})$. This minor distinction between matrices and the Möbius transformations is mildly confusing when it comes to counting the relative size of groups.

In the days before the "new maths" that eventually followed Sputnik one first met matrices in the context of linear fractional transformations. The use of matrices to confuse undergraduates trying to solve systems of linear equations is a fairly modern phenomenon.

VII.9 I don't quite know why mathematics becomes "maths" in Australian English, whereas it is "math" in American.* This wouldn't be any more important than the many other differences in English and American spelling, were it not for the opportunity it presents to have one's e-mail bounce.

---

*My advisers (specifically, Victor Scharaschkin) point out that this likely is a case of conservation of "s"s. The missing "s" reappears redundantly in a variety of locutions.

VII.10 Typically, a $'$ on a summation sign tells not to include some term which obviously should not be included. I think of it as an instruction not to be silly.

VII.11 I may come to this again later, but let me say here that the function G_{2k} is said to be a modular form of *weight* $2k$. Worse actually, several writers call it a modular form of weight k. And in some older books instead, its "dimension" is $-2k$. I have difficulty in resisting the belief that such things are done to ensure that only the *cognoscenti* become experts in the field. Still, I'll probably accidentally perpetuate some like ambiguity. Anyhow, the unambiguous remark I need to make is that a modular form of weight 0, thus a modular form that is honest-to-goodness *invariant* under some appropriate group, is called a modular *function*.

VII.12 My being lazy and refusing to calculate that simple constant was because I felt threatened by having to use that unpleasant differential equation to do the work. I should have looked ahead to Lecture X to learn that the Fourier expansions report that near infinity one has

$$G_4 = \frac{\pi^4}{3^2 \cdot 5}(1 + \cdots) \qquad G_6 = \frac{2\pi^6}{3^3 \cdot 5 \cdot 7}(1 + \cdots),$$
$$G_8 = \frac{\pi^8}{3^3 \cdot 5^2 \cdot 7}(1 + \cdots), \quad \text{and} \quad G_{10} = \frac{2\pi^{10}}{3^5 \cdot 5 \cdot 7 \cdot 11}(1 + \cdots).$$

It is now trivial to find that

$$G_4^2 = \tfrac{7}{3}G_8 \quad \text{and} \quad G_4G_6 = \tfrac{11}{5}G_{10}.$$

VII.13 Frey first mentioned his notions at an Oberwolfach meeting, and subsequently reported in detail in a local journal ['Links between stable elliptic curves and certain diophantine equations', *Ann. Univ. Sarav. Math. Ser.* **1** (1986), pp. 1–40]; a later accessible reference is Gerhard Frey, 'Links between solutions of $A - B = C$ and elliptic curves', in H. P. Schlickewei and E. Wirsing eds., *Number Theory, Ulm 1987* (15th Journées Arithmétiques, Ulm 1987), Springer Lecture Notes **1372** (New York: Springer–Verlag, 1989). But Frey's remarks were not the first to suggest a connection between elliptic curves and Fermat's Last Theorem. At Besançon in 1972, Yves Hellegouarch wrote a thesis 'Courbes elliptiques et équation de Fermat'. However, Hellegouarch applies known information about the FLT to elliptic curves rather than using the elliptic curves to learn about Fermat's equation; see 'Points d'ordre $2p^h$ sur les courbes elliptiques' [*Acta Arith.* **26** (1975), pp. 252–263].

VII.14 Frey has told me that I am helping to perpetuate a legend with the above remark. Frey had studied his curves earlier independently of, but in exactly the context of, the work of Hellegouarch. He points out that his observations at Oberwolfach were not just a random aside, but a natural development of his work. Indeed, the earlier ideas had arisen in the context of studying Ogg's conjecture (later to become Mazur's theorem) on torsion points on rational elliptic curves. At that time the connection to Fermat's Last Theorem was a bad thing, an obstruction. After all, one knew the FLT to be inaccessible, so it was never good to find it intruding into matters one hoped to prove. On the other hand, after Mazur's work it became reasonable to ask whether knowledge of elliptic curves might yield insight into the FLT.

Gerhard Frey at the Boston meeting, August 1995

Photograph by C. J. Mozzochi, Princeton

LECTURE VIII

It's surely reasonable to say that if something is true locally everywhere, then it is true globally, and conversely. What's not so clear is what one might mean in saying such a thing in the context of diophantine problems.

Let me make some suggestions, starting with the Riemann ζ-function $\zeta(s) = \sum n^{-s}$. By the way, here s is a complex variable, traditionally given as $s = \sigma + it$. Of course, the series converges absolutely only in the right plane $\sigma > 1$. We owe to Euler the "factorization"

$$\sum_{n=1}^{\infty} \frac{1}{n^s} = \prod_p \left(1 - \frac{1}{p^s}\right)^{-1}.$$

Here the product is over all primes $p = 2, 3, 5, \ldots$. The Euler factor is $1 + p^{-s} + p^{-2s} + p^{-3s} + \cdots$, and the product representation amounts to the unique factorization theorem whereby each positive integer has a unique representation as a product of primes.

In this case we might hope to obtain local information, that is about the primes, from global knowledge of the positive integers. For example, because $1 + \frac{1}{2} + \frac{1}{3} + \cdots + \frac{1}{n} \sim \log n$ we find, after taking a logarithm, that $\sum_{p<x} p^{-1} \sim \log\log x$. Since $\log\log x = \int_e^x (t \log t)^{-1} dt$, we might deduce that the m th prime is $\sim m \log m$, and therefore that the number of primes less than x is $\sim x/\log x$.

But I have something more down-to-earth in mind. Suppose that we're wondering whether there are rational points on the curve $x^4 - 17 = 2y^2$ of genus 1. First we'll set $x = X/Z$, $y = Y/Z$ and multiply by Z^4 to obtain a homogeneous polynomial in the *projective* coordinates X, Y, and Z — this also has the advantage of giving a glimpse of ∞: just consider the line $Z = 0$. But, whatever, after failing to notice a solution we might well choose to consider the problem modulo p — with the intention of using the evident principle that if there is no solution modulo m, then there is certainly none *globally*; that is, there then cannot be a rational solution.

However, in the present example there is a solution mod p for all p. That doesn't require an infinite amount of work; in fact, it suffices to check by trial and error at $p = 2$ and 17, the other *places* being quite automatic. Of course, we want to deal with an arbitrary modulus m, so we also need the powers of the primes. However once one has a solution modulo p, it is easy to see whether or not there is a solution modulo p^n for all n. Let me illustrate that by showing that $x_{n-1}^2 \equiv 2 \pmod{7^n}$ has a solution for all positive integers n, given that $3^2 \equiv 2 \pmod 7$. It will give a feeling of generality if I set $f(x) = x^2 - 2$ and $p = 7$.

Suppose we already know that $x_0 = a_0, x_1, x_2, \dots$, with

$$x_{n-1} = a_0 + a_1 p + a_2 p^2 + \cdots + a_{n-1} p^{n-1}$$

solving $f(x_{n-1}) \equiv 0 \pmod{p^n}$. We next need $x_n = x_{n-1} + a_n p^n$ so that $f(x_{n-1} + a_n p^n) \equiv 0 \pmod{p^{n+1}}$. Set $a_n = x$. The Taylor expansion

$$f(x_n) = f(x_{n-1} + xp^n) = f(x_{n-1}) + xp^n f'(x_{n-1}) + \cdots ,$$

where the $\cdots$ indicate terms in yet higher powers of p, tells us that

$$0 \equiv p^{-n} f(x_{n-1}) + xf'(x_{n-1}) \equiv p^{-n} f(x_{n-1}) + xf'(x_0) \pmod p .$$

Recall that $p^n \mid f(x_{n-1})$ and $x_{n-1} \equiv x_0 \pmod p$. So we have only to solve a simple linear congruence in x to obtain x_n. This argument is known as Hensel's lemma. Mind you, there is a little more work to be done if x_0 happens to be a singular point, that is if $f'(x_0) \equiv 0$, but ultimately things settle down to just a sequence of linear problems as above. So it's almost painless to decide whether there is a solution modulo p^∞, so to speak. In our example we obtain

$$x_\infty = 3 + 1 \cdot 7 + 2 \cdot 7^2 + 6 \cdot 7^3 + 1 \cdot 7^4 + 2 \cdot 7^5 + 1 \cdot 7^6 + 2 \cdot 7^7 + 4 \cdot 7^8 + \cdots .$$

It is a well-known principle both of magic and of mathematics that once one knows the true name of an object, one controls it. Thus we say that x_∞ is a 7*-adic integer*. We observe that we can do arithmetic with objects such as it, much as if they were absolutely convergent power series in powers of 7, provided that we remember to "carry" as the means of coping with coefficients outside the range $\{0, 1, 2, 3, 4, 5, 6\}$.

To see what is actually going on here it is best first to remember that we barely understand the real numbers $\mathbb{R}$. We would be satisfied to say that the reals are the set of all decimals, and proud to pop out with the statement that "$\mathbb{R}$ is the field of limit points of convergent sequences of rationals," which it is. If asked to clarify this claim, we'd remark that a sequence $(a_h) = \{a_0, a_1, a_2, \dots\}$ of rationals is convergent if it is Cauchy; that is, if $|a_n - a_m|$ is arbitrarily small as soon as m and n are sufficiently large, we'd show that the termwise difference and product of Cauchy sequences is again Cauchy. We'd agree to identify such sequences if they have the same limit — that is, if they differ by a *null sequence*: one for which $|a_n|$ is as small as we like whenever n is large enough. In a trice we would have taken a leap in sophistication and would hear ourselves saying that "$\mathbb{R}$ is just the ring of Cauchy sequences of rationals modulo its maximal ideal of null sequences."

The joker in this explanation is that we could give it without telling anyone what we meant by "small", that is, what we mean by the absolute value $|\ \ |$. Well, yes, we'd have used that $|\ \ |$ is a positive definite map on the rationals preserving multiplication — $|a| \geq 0$ and $= 0$ if and only if $a = 0$; and $|a \cdot b| = |a| \cdot |b|$. And we will have needed the *triangle inequality*: $|a + b| \leq |a| + |b|$. But that's it.

So a listener could say, "Aha! 'small' means 'divisible by a high power of my favorite prime p'. Having picked p I could take any rational r/s and set $|r/s| = p^{-\nu_p}$." Here she would have applied unique factorization to factorize $r/s = \prod_p p^{\nu_p}$. Most of the ν_p would be zero, the rest positive or negative integers.

After making her write $|\ \ |_p$ instead of just $|\ \ |$ — we might agree to write $|\ \ |_\infty$ for the absolute value we thought we had meant, so as not to make too special a claim for it — we'd find that all was well. She has indeed nominated an absolute value and her *completion* $\mathbb{Q}_p$ — the p-adic rationals — has as much right to respect as does our real completion $\mathbb{R}$.

Is $\mathbb{R}$ special? Well, physical reality, as we and Archimedes believe it, has no infinitesimals. In $\mathbb{R}$, given $a \neq 0$ no matter how close to 0, and b no matter how large, there is always an integer n sufficiently large that $|na|_\infty > |b|_\infty$. Applied just to the integers and the usual absolute value, this is the Archimedean Axiom. Hence we say that the field $\mathbb{R}$ and its absolute value $|\ \ |_\infty$ are *archimedean*. However, the new absolute values are quite different. The p-adic absolute values satisfy the *ultrametric* inequality $|a + b|_p \leq \max(|a|_p, |b|_p)$, a stronger form of the triangle inequality. That entails $|na|_p \leq |a|_p$ for all integers n; so integers are never "large". We say that the fields $\mathbb{Q}_p$ and thence the absolute values $|\ \ |_p$ are *nonarchimedean*. By the way, and *pace* Archimedes, in recent years the theoretical physicists have learned about the p-adic fields and have begun to wonder whether they may not help in modeling just what happens in the nuclei of atoms, for instance. So much for reality.

The x_∞ of our example is an element of the field $\mathbb{Q}_7$. It is in fact a 7-adic *integer*, that is, it satisfies $|x_\infty|_7 \leq 1$. Just as $\mathbb{Q}$ is the field of quotients of elements of $\mathbb{Z}$, so the fields $\mathbb{Q}_p$ are the respective quotient fields of the domains $\mathbb{Z}_p$. The elements of $\mathbb{Q}_p$ can be represented as sums $\sum_{n=-m} a_n p^n$ just as the elements of $\mathbb{R}$ are often represented as decimals $\sum_{n=-m} b_n 10^{-n}$. All's well, because just as 10^{-n} gets small at exponential rate in $\mathbb{R}$, so p^n becomes small in $\mathbb{Q}_p$.

The bottom line is that checking diophantine equations modulo m can usefully, and elegantly, be viewed as asking for solutions in fields $\mathbb{Q}_p$ of p-adic rationals. That is true, using the Chinese Remainder Theorem, whereby any congruence modulo $m = \prod p^{\operatorname{ord}_p m}$ is the same thing as a collection of congruences modulo the respective $p^{\operatorname{ord}_p m}$. One talks about looking for solutions *at* p; these are the solutions *locally*. But we'd better not forget about ∞. The diophantine equation $x^2 + y^2 = -1$ has solutions at all p, but it really has no rational solutions, because it has no solutions in $\mathbb{R}$. Thus when we say *locally everywhere* we mean at all p *and* at ∞.

The distinct absolute values are quite different; they measure smallness in completely incompatible ways. But they are not totally unrelated. For nonzero x, unique factorization gives the "product formula"

$$\sum_p \log |x|_p = -\log |x|_\infty .$$

With all that said, is it true that if a diophantine equation has solutions locally everywhere — thus in all $\mathbb{Q}_p$ and in $\mathbb{R}$ — then it has solutions in $\mathbb{Q}$ — thus a *global* solution? Sadly, no.

There is a theorem of Hasse-Minkowski showing that a *quadratic form* — a homogeneous polynomial expression of degree 2 — takes a given rational value if and only if it the resulting equation has a solution locally everywhere. But this *Hasse principle* does not apply generally. In fact, I obnoxiously chose the example equation $x^4 - 17 = 2y^2$ to display an equation with solutions locally everywhere but with no rational solutions. Nonetheless, our view of diophantine equations is that they would like to obey the Hasse principle. Accordingly, one studies the obstruction to their doing so. The role of the Tate-Shafarevitch group Ш, of which we'll hear later, is to detail that obstruction in the case of curves of genus 1.

An elementary aside. Let $r/s \in \mathbb{Q}$ be a rational zero of some polynomial $f(x) = a_0x^n + a_1x^{n-1} + \cdots + a_{n-1}x + a_n \in \mathbb{Z}[x]$. If $p \mid r$, then this zero is 0 (mod p), so 0 must be a zero of $f(x)$ (mod p). That is, we have $a_n \equiv 0 \pmod{p}$, which means $p \mid a_n$. This beginning of a local argument of course eventually entails that $r \mid a_n$. Now consider the *reciprocal* polynomial $x^n f(x^{-1})$. We see it has s/r as a zero and conclude that $s \mid a_0$.

In the particular case when $f(x)$ happens to have 0 as a zero we have an instructive *degenerate* case of the preceding discussion. Since everything divides 0, anything whatsoever must divide a_n, showing that of course $a_n = 0$. Now the preceding argument would have us say that the reciprocal polynomial has ∞ as a zero. Indeed, by rights it should be a polynomial of degree n. The reciprocal polynomial displays its zero at ∞ by actually being of lower degree than n.

A generalization of my earlier remark is Gauss's lemma. It asserts that if a polynomial $f(x)$ with integer coefficients has a factorization as a product of polynomials with rational coefficients, then it already has a factorization into a product of polynomials with integer coefficients.

To see this we write $df(x) = g(x)h(x)$, where the nonzero integer d is a multiplier required so that g and h have integer coefficients. If we fear that $p \mid d$, we look at the identity modulo p, obtaining $0 \equiv g(x)h(x)$ modulo p. But with p prime, this entails that modulo p, either $g(x) \equiv 0$ or $h(x) \equiv 0$. So we may divide through by p, in particular replacing d by d/p. By descent we may conclude that d has no prime divisors. That is $d = 1$, as we wished to show.

The terminology "global" and "local", and the talk of "at" p is sometimes jolting. Its genesis can readily be illustrated. Suppose we think about the arithmetic of functions $f(z)$ defined on $\mathbb{C}$. It seems reasonable to consider the polynomials as the integers and the *rational functions* — quotients of polynomials — as the rationals. What are the irreducibles? That's easy, they're just the linear polynomials $z - \alpha$, one for each point $\alpha \in \mathbb{C}$. So that's why we say "at" a prime! To say that $z - \alpha$ divides some polynomial f exactly $m = \text{ord}_\alpha f$ times is to say that f has a zero of order $\text{ord}_\alpha f$ at α; or, if you prefer, that the Taylor expansion of f at α has $(m!)^{-1} f^{(m)}(\alpha)(z - \alpha)^m$ as its first nonzero term.

And just as the numbers have a special "prime" corresponding to the absolute value $| \quad |_\infty$, so there morally is a place in $\mathbb{C}$, the point* at ∞.

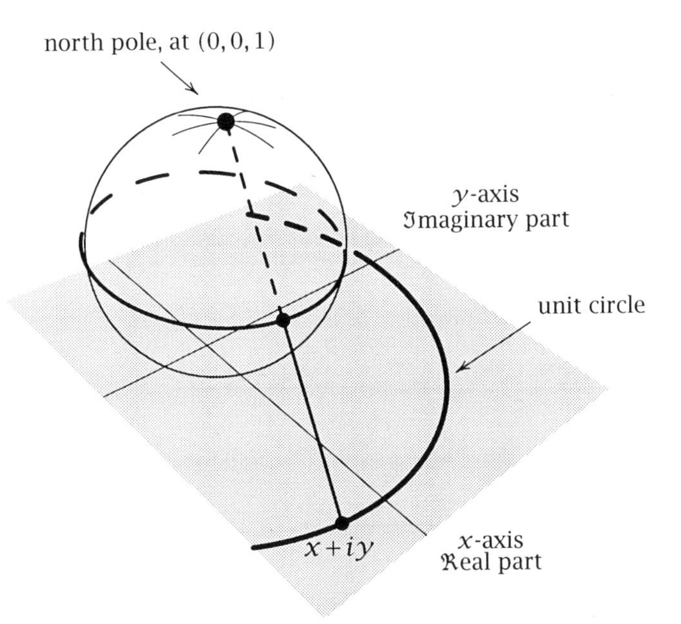

Riemann sphere, radius $\frac{1}{2}$; equator maps to unit circle

The fundamental theorem of algebra that, in $\mathbb{C}$, a polynomial f of degree $\deg f$ has exactly $\deg f$ zeros counted according to multiplicity, becomes the formula $\sum_{\alpha \in \mathbb{C}} \text{ord}_\alpha f = -\text{ord}_\infty f$.

Geometry with an ultrametric is great fun. Every triangle is isosceles and each point inside a circle is its center. However, we're interested in arithmetic and what pleases us is that in $\mathbb{Q}_p$ the integers $\mathbb{Z}$ no longer discretely keep their distance one from the other.

*Plainly, $\mathbb{C}$ is not plane, but spherical. Just put a globe on the origin of the plane and draw a line from the north pole N to each point in $\mathbb{C}$. Now identify each point in $\mathbb{C}$ with the point on the globe cut by its line. Clearly, N corresponds to ∞, and we may speak of *the* point at ∞.

We can then look at difficult functions like the Riemann ζ-function in a new light. After all, we are a bit blasé to take a function like n^{-s} for granted. An honest integral power, like n^{-2k}, is fine; it just means n^{-2} multiplied by itself k times. But suddenly to generalize that to an analytic function $e^{-s\log n}$ is a drastic step. In that spirit, one might well feel that, except for the perfectly decent rational numbers $\zeta(2k)/\pi^{2k}$, all the other $\zeta(s)$ should be disregarded. Yet it was nice to have an analytic function. Now a miracle comes to the rescue. Kummer's congruences: if

$$k \equiv k' \pmod{p^n} \quad \text{then} \quad \frac{(1-p^{k-1})B_k}{k} \equiv \frac{(1-p^{k'-1})B_{k'}}{k'} \pmod{p^{n+1}}$$

amount exactly to our being able to view the numbers

$$(-1)^k(1-p^{2k-1})(2k-1)!\,\frac{\zeta(2k)}{2^{2k-1}\pi^{2k}}$$

as the values of a p-adic analytic function of a p-adic variable k. This is the p-adic ζ-function.

Notes and Remarks

VIII.1 One of my advisers cruelly suggests that the eccentric standard notation $s = \sigma + it$ is a consequence of Riemann not knowing enough Greek to be familiar with the letter τ.

VIII.2 It does take a moment to see that

$$\prod_p (1 + p^{-s} + p^{-2s} + p^{-3s} + \cdots) = \sum_{n=1}^{\infty} \frac{1}{n^s}$$

is just the unique factorization theorem for the positive integers. In a similar spirit Euler also observed that

$$\prod_{n=0}^{\infty} (1 + x^{2^n}) = 1 + x + x^2 + x^3 + x^4 + x^5 + \cdots = \frac{1}{1-x}.$$

Equivalently, each positive integer has a unique binary representation.

VIII.3 It is perfectly sensible not to want to have anything to do with infinite products. Fortunately, the simple act of taking their logarithm turns them into relatively harmless infinite sums. That's moreover what the theory amounts to. One defines infinite products by

$$\log\Big(\prod_{n=0}^{\infty}(1-u_n)\Big) = -\sum_{n=0}^{\infty}\Big(u_n + \tfrac{1}{2}u_n^2 + \tfrac{1}{3}u_n^3 + \cdots + \tfrac{1}{m}u_n^m + \cdots\Big)$$

and remarks that it is plainly necessary for absolute convergence on the right that the sum $\sum_{n=0}^{\infty} |u_n|$ converges. A moment's further thought now shows that this is already sufficient for absolute convergence on the right, so one concludes that the infinite product $\prod_{n=0}^{\infty}(1+u_n)$ converges absolutely if and only if the sum $\sum_{n=0}^{\infty} u_n$ converges absolutely.

VIII.4 Given what I've said about primes and the Riemann ζ-function, it is natural to study the primes by considering, for example,

$$\frac{1}{\zeta(s)} = \prod_p \left(1 - \frac{1}{p^s}\right).$$

The point is to provide error estimates for the crude guesses I cited. The first important thing is to show that the reciprocal is relatively harmless at $s = 1$; the product diverges to 0. With that one has taken a long stride towards the *Prime Number Theorem* of Hadamard and de La Vallée-Poussin, whereby indeed, if $\pi(x)$ denotes the number of primes less than x, then $\lim_{x\to\infty} \pi(x) \,/\, \frac{x}{\log x} = 1$.

VIII.5 But let's look more closely at the product $\prod_p(1-p^{-s})$. Multiplying out, we get the sum $\sum \mu(n)n^{-s}$, where the *Möbius* function $\mu(n)$ is just $(-1)^{\nu(n)}$ if n has $\nu(n)$ different prime factors and is squarefree, whereas $\mu(n) = 0$ if n is divisible by a nontrivial square. At first glance the Möbius function seems dull and contrived. Yet it deserves serious respect because of its property

$$\sum_{d|n} \mu(d) = 0 \quad \text{if} \quad n > 1; \quad \text{and} \quad = 1 \quad \text{if} \quad n = 1.$$

Here the sum is over all positive integer divisors d of n.

To see why this property of the Möbius function is important, we notice the manner in which *Dirichlet series* $\sum_{n=1}^{\infty} a_n n^{-s}$ multiply. The rule is fairly easily checked to be that if

$$\sum_{n=1}^{\infty} a_n n^{-s} \sum_{n=1}^{\infty} b_n n^{-s} = \sum_{n=1}^{\infty} c_n n^{-s} \quad \text{then} \quad c_n = \sum_{d|n} a_n b_{n/d}.$$

It is commonplace in arithmetic to want to study functions F of the shape

$$F(n) = \sum_{d|n} f(d).$$

By what was just remarked we have that

$$\sum_{n=1}^{\infty} F(n)n^{-s} = \sum_{n=1}^{\infty}\Big(\sum_{d|n} 1 \cdot f(d)\Big)n^{-s} = \zeta(s)\sum_{n=1}^{\infty} f(n)n^{-s}.$$

Hence $\sum_{n=1}^{\infty} f(n)n^{-s} = (\zeta(s))^{-1}\sum_{n=1}^{\infty} f(n)n^{-s}$ and it follows that

$$f(n) = \sum_{d|n} \mu(n/d)F(d).$$

This is known as Möbius inversion. It is amusing just how often Möbius inversion — of course in heavy disguise — is rediscovered in other parts of mathematics. Among the more popular arithmetic functions one has the divisor functions $\sigma_k(n)$ denoting the sum of the kth powers of the divisors of n.

But the function I want to mention again is the Euler totient function $\phi(n)$, which counts the positive integers less than, and prime to n. We met it in the notes to Lecture III in the context of Euler's generalization of Fermat's theorem and were told that $\sum_{d|n}\phi(d) = n$. By the way, that's clear because, obviously, $\sum_{d|n}\#\{k : \gcd(k,n) = d;\ 1 \le k \le n\} = n$ and, plainly, $\#\{k : \gcd(k,n) = d;\ 1 \le k \le n\} = \phi(n/d)$.

Then by Möbius inversion we have

$$\phi(n) = \sum_{d|n}\mu(d)n/d = n\sum_{d|n}\mu(d)/d = n\prod_{p|n}\left(1 - \frac{1}{p}\right).$$

Here the product is over all positive integer primes p dividing n. By the way, one often sees "Moebius" rather than "Möbius". This is not bad spelling. It is *de rigueur* in German to replace that Umlaut by a trailing "e" in contexts where one fears that the Umlaut might not get printed.

VIII.6 The Chinese Remainder Theorem is just the statement

孫子定理

若 $m_1, m_2, \dots, m_n$ 互素，則

$$x \equiv a_i \pmod{m_i}, \quad 1 \le i \le n$$

有唯一解，mod $m_1 \cdots m_n$.

You will have noticed that the theorem is attributed to Sun Tzu, the author of *The Art of War*, suggesting the result is some two thousand years old.

VIII.7 What can I have meant by suggesting that it is *quite automatic* that $x^4 - 17 = 2y^2$ has solutions modulo p, for all primes p? Of course, I just quoted from a source: J. W. S. Cassels, 'Diophantine equations with special reference to elliptic curves' [*J. London Math. Soc.* **41** (1966), pp. 193–291]. However, we can indeed deal with all the primes in just a few bites. First, a nibble. The case $p = 2$ has the solution $x = 1$, $y = 0$. Onto the big bites: If $p \equiv 3 \pmod 4$, then $x^4 - 17$ takes on $\frac{1}{2}(p-1) + 1$ different values modulo p, as does $2y^2$; so there must be a common value. The point is that there are as many fourth powers as there are squares mod p because all the squares are fourth powers. This is not so if $p \equiv 1 \pmod 4$. However, if $p \equiv 1 \pmod 8$, then -2 is a square modulo p whence $x = 1$

provides a solution. On the other hand, if $p \equiv 5 \pmod 8$, then -2 is not a square modulo p. Suppose that 17 is also not a square modulo p. Then $-17/2$ *is* a square, so $x = 0$ provides a solution. Finally, suppose that 17 is a square modulo p. Either 17 is a fourth power, in which case $y = 0$ gives an easy solution, or -17 is a fourth power whence $x \equiv \sqrt[4]{-17}$ and $y \equiv x^2$ satisfies the equation. Yup, quite automatic, as you now see. The case $p = 17$ needs a second glance lest it provides the solution $(0,0)$, obstructing easy use of Hensel's lemma; in practice there is no problem.

VIII.8 A p-adic number is just an object

$$\alpha = a_m p^{-m} + \cdots + a_{-1}p^{-1} + a_0 + a_1 p + a_2 p^2 + \cdots ,$$

where the "digits" a_i belong to $\{0, 1 \ldots , p - 1\}$. Such numbers look like formal Laurent series in p. But one does arithmetic with these objects much as one does with decimals, with all the problems of "carrying" to keep the digits restricted. Notice that $|\alpha|_p = p^m$, of course supposing that indeed $a_{-m} \neq 0$. The p-adic rational α is said to be an *integer* if $a_j = 0$ for $j < 0$; equivalently, if $|\alpha|_p \leq 1$. The p-adic integers are invariably denoted $\mathbb{Z}_p$ and the p-adic rationals by $\mathbb{Q}_p$. Of course, the ordinary "rational" integers $\mathbb{Z}$ belong to $\mathbb{Z}_p$ for each p. That has the congenial effect that, p-adically, any infinite collection of elements of $\mathbb{Z}$ has a p-adic cluster point; that is, $\mathbb{Z}$ is a compact subset of the compact set $\mathbb{Z}_p$.

VIII.9 My remarks on p-adic numbers in the Lecture truly are horribly compacted. Perhaps I can do better by repeating *holus bolus* a section* from 'Some problems concerning recurrence sequences', by G. Myerson and A. J. van der Poorten [*Amer. Math. Monthly* **102** (1995), pp. 698–705]:

The absolute value function defined on the integers has the following properties;

(i) $|x| \geq 0$ for all x;
(ii) $|x| = 0$ if and only if $x = 0$;
(iii) $|xy| = |x| \cdot |y|$ for all x and y; and
(iv) $|x + y| \leq |x| + |y|$ for all x and y.

There are other functions that have the same properties. Given any non zero integer n and any prime number p, we can write $n = p^a m$ with a and m integers, $a \geq 0$ and m relatively prime to p. Moreover, this expression is unique. Define the function $|\ |_p$ by $|n|_p = p^{-a}$. Thus for example, $|35|_7 = \frac{1}{7}$, $|36|_7 = 1$, and $|36|_3 = \frac{1}{9}$. If by convention we take $|0|_p = 0$ for all p, then it is not hard to see that all the properties of $|\ |$ listed above hold for $|\ |_p$, for each p. In fact, the last property holds in a stronger form, namely

(iv′) $|x + y|_p \leq \max(|x|_p, |y|_p)$.

---

*Repeated here with the permission of at least one of the authors. [Of course, Gerry was enthusiastic about my including the material.]

We call $|\ \ |_p$ the *p-adic absolute value.* Thinking about convergence with respect to this absolute value leads to some peculiar-looking formulas. For example, for the geometric series with first term 6 and common ratio 7, the equation

$$6 + 42 + 294 + 2058 + \cdots = -1$$

is a blunder in the usual run of things, but quite correct in the 7-adics.

The p-adic absolute value is easily continued to a function on usual rational numbers, enjoying properties (i) through (iv$'$); any rational x can be written as $x = p^a \frac{r}{s}$ for a, r, s integers with r and s both relatively prime to p. Thus $|\frac{35}{36}|_7 = \frac{1}{7}$ and $|\frac{35}{36}|_3 = 9$.

Any rational x has a unique decimal expansion $x = \sum_{j=m}^{\infty} a_j 10^{-j}$ with a_j in $\{0, 1, \ldots, 9\}$, the series converging in the usual absolute value. So too for each p, any rational x has unique p-adic expansion of the form $x = \sum_{j=m}^{\infty} a_j p^j$, with a_j in $\{0, 1, \ldots, p-1\}$, converging in the p-adic absolute value. For example, because

$$\frac{1}{1-7} = 1 + 7 + 7^2 + \cdots$$

in the 7-adics we have

$$\begin{aligned} \frac{17}{98} &= 7^{-2} \cdot \frac{17}{2} = 7^{-2}\left(9 + \frac{3}{1-7}\right) \\ &= 7^{-2}(2 + 1 \cdot 7 + 3 + 3 \cdot 7 + 3 \cdot 7^2 + \cdots) \\ &= 5 \cdot 7^{-2} + 4 \cdot 7^{-1} + 3 + 3 \cdot 7 + 3 \cdot 7^2 + \cdots . \end{aligned}$$

Now consider the sequence 1, 1.4, 1.41, 1.414, 1.4142, ... of decimal approximations to the square root of 2. If m is less than n, then the mth and nth terms of this sequence differ by less than 10^{-m}, a quantity which goes to zero as m increases. Such a sequence is called a *Cauchy sequence* (with respect to the usual absolute value).

You can't help feeling such a sequence ought to have a limit, but this one doesn't — *if* you confine yourself to the rationals [Euclid, *The Thirteen Books of Euclid's Elements* 2nd ed., T. L. Heath, ed. (New York: Dover, 1956)]. In analysis, it is useful for Cauchy sequences to have limits, so we embed the rationals in the larger set called the *reals.* Every real number has a decimal expansion, and every Cauchy sequence converges—we say the reals are *complete.* The details of the completion process can be found in many introductory analysis texts, for example A. Gleason, *Fundamentals of Abstract Analysis*, (Reading, Mass.: Addison-Wesley, 1966).

Now consider the sequence: $7, 7 + 7^2, 7 + 7^2 + 7^4, 7 + 7^2 + 7^4 + 7^8, \ldots$. In 7-adic absolute value, the difference between the mth and nth terms in this sequence is $|7^{2^m} + \cdots + 7^{2^{n-1}}|_7 = 7^{-2^m}$, which clearly goes to zero as m increases. That is to say, this is a Cauchy sequence — *if* you view it 7-adically. It ought then to have a limit. It is not a geometric series, so it cannot have a rational limit. By a process formally identical to the construction of the reals, we embed the rationals in a larger set we denote $\mathbb{Q}_p$, called *the p-adic rationals.* Every p-adic rational has a p-adic expansion and the p-adic rationals are complete.

Back to the reals. There are nonconstant polynomials which have real coefficients but no real roots, for example $x^2 + 1$. If we extend the reals to a field containing a root of $x^2 + 1$, we obtain the complex numbers. *Mirabile dictu,* every nonconstant polynomial with complex coefficients has a complex root. We say that the complex numbers are *algebraically closed.* The absolute value function is continued to the complex numbers by $|a + bi| = (a^2 + b^2)^{1/2}$. *Mirabile* squared, the complex numbers are complete (with respect to this absolute value). The important functions of calculus (rational, exponential, trigonometric, ...) can be continued to functions of a complex variable and many problems about real functions become easier to handle in this larger domain.

Back to the p-adic rationals. They are not algebraically closed. For example, if α in $\mathbb{Q}_7$ were a root of $x^2 - 7 = 0$, we would have $|\alpha|_7 = 7^{-1/2}$. But if α had 7-adic expansion $\alpha = \sum_{j=m}^{\infty} a_j 7^j$, we would have $|\alpha|_7 = 7^{-m}$, with m an integer. We can embed $\mathbb{Q}_p$ inside an algebraically closed field $\overline{\mathbb{Q}}_p$, although the miracle of "add one number, get the rest free" does not occur here. We can extend $|\ \ |_p$ to $\overline{\mathbb{Q}}_p$, but $\overline{\mathbb{Q}}_p$ is not complete. We can complete $\overline{\mathbb{Q}}_p$ to a field $\mathbb{C}_p$, which is the p-adic analogue of complex numbers; it is complete and algebraically closed. There is a rich theory of analytic functions on $\mathbb{C}_p$, mirroring that on the complex numbers. This material can be found in less telegraphic form in N. Koblitz, *p-adic Numbers, p-adic Analysis, and Zeta-Functions* (New York, Springer–Verlag, 1977).

What is really going on is this: The set of all Cauchy sequences forms a ring once we define the operations termwise; that the set is closed under the operations is a consequence of the rules (i)–(iv). One defines the field of reals (respectively, p-adic rationals, according to the particular valuation defining "Cauchy") to be this ring with sequences "with the same limit" identified. What that means is that we take the subset of null sequences, those converging to 0, and notice again by the rules (i)–(iv) that this set is a maximal ideal in the ring of Cauchy sequences. Then the quotient ring is a field.

The "miracle" of $\mathbb{R}$ and $\mathbb{C}$ actually is rather special. It turns out that if a field $\mathbb{F}$ is algebraically closed and if $\mathbb{L}$ is a subfield of finite codimension in $\mathbb{F}$ (in English: if $\mathbb{F}$ is a finite-dimensional vector space over a field $\mathbb{L}$), then necessarily $[\mathbb{F} : \mathbb{L}] = 2$ (compare $[\mathbb{C} : \mathbb{R}] = 2$) and $\mathbb{L}$ is an *ordered field.* That means that $\mathbb{L}$ is the disjoint union of three sets N, $\{0\}$, P with P closed under addition and multiplication, such that $N = -P$; P is of course the set of *positive* elements of $\mathbb{L}$. It turns out that $\mathbb{L}$ can be ordered if and only if -1 is not a sum of squares. A complete orderable field is known as a *real field* and always is a subfield of codimension 2 in an algebraically closed field; for all this see, for example, [Serge Lang, *Algebra* (Reading,

Mass.: Addison-Wesley, 1965), Chapter XI. By contrast, it is not hard to see that $\mathbb{Q}_p$ is not an ordered field.

VIII.10 Given that closing paragraph above it seems worthwhile to mention some details on just why the "miracle squared" that $\mathbb{C}$ is algebraically closed. Remember, this is the "fundamental theorem of algebra". Suppose it is false; in other words, suppose that there is a polynomial of degree $m > 1$ over $\mathbb{C}$ that is irreducible over $\mathbb{C}$.

Then there is an extension field of degree m over $\mathbb{C}$, and thus of degree $2m$ over $\mathbb{R}$. Suppose that m has an odd prime factor p. Any group with order divisible by p has a normal subgroup of index p. By galois theory there then is a field extension of $\mathbb{R}$ of odd degree. It must be generated by a zero of an irreducible polynomial of odd degree over $\mathbb{R}$. But that's absurd. Every polynomial of odd degree over $\mathbb{R}$ has a zero in $\mathbb{R}$. We know that from graphing polynomials. So m must be a power of 2. But, again by galois theory, this entails that there is an extension of degree 2 of $\mathbb{C}$. However, — this takes a moment's work to verify — every complex number has a complex square root. So $\mathbb{C}$ cannot have an extension of degree 2 and so m must be 1.

VIII.11 While it seems natural to use the elements of $\{0, 1 \dots, p-1\}$ as the p digits when detailing a p-adic number, it is often more useful to use, instead, the *Teichmüller symbols* 0 and the $p-1$ different $(p-1)$ th roots of unity. That works because it follows from Fermat's Little Theorem that, modulo p, those $p-1$ roots are respectively congruent to $\{1, 2 \dots, p-1\}$.

At this point I acquired the sudden urge to give an actual example, and computed the symbols for $p = 7$. Seven is not too large, I said to myself, but is nontrivial. Of course, $\omega(0) = 0$, $\omega(1) = 1$, and

$$\omega(2) = 2 + 4 \cdot 7 + 6 \cdot 7^2 + 3 \cdot 7^3 + 0 \cdot 7^4 + 2 \cdot 7^5 + 6 \cdot 7^6 + 2 \cdot 7^7 + \cdots .$$

Naturally, $\omega(6) = -1$, $\omega(5) = -\omega(2)$, and $\omega(4) = -\omega(3)$, but I was at first incredulous to find the program PARI allege that $\omega(3) = \omega(2) + 1$. So much so that I complained to Henri Cohen, PARI's author, that his program had broken. My signature, at the time, happened to include the wise quotation: "Genius may have its limitations, but stupidity is not thus handicapped." Indeed. I should have noted its advice.

What was up is that it *is* so. I was just computing the 7-adic sixth roots of unity. The construction of a regular hexagon is easy precisely because its edges have the same length as the radius of the circumscribing circle. In fact, $\phi(6) = 2$ should have reminded me that there really is only one pair of conjugate irrationalities involved.

VIII.12 The PARI manual explains that "The PARI system is a package which is capable of doing formal computation on recursive types at high speed; it is primarily aimed at number theorists but can be used by people whose primary need is speed." The program was created by a group at

Bordeaux and is available free to users.* For the mathematics involved and much more see Henri Cohen's book, *A Course in Computational Algebraic Number Theory*, Graduate Texts in Mathematics **138** (New York: Springer–Verlag, 1993). It's a very good book.

VIII.13 I'm pushing my luck in trying to say all that I have said about p-adic numbers in just a few pages. You may do better by reading the very useful elementary introduction provided by Kurt Mahler, *p-Adic Numbers and their Functions*, Cambridge Tracts in Mathematics **76**, (Cambridge: Cambridge University Press, 1981); here one gets an opportunity to read a master and an introduction at one and the same time. Two good books: *p-adic Numbers, p-adic Analysis, and Zeta-Functions*, Graduate Texts in Mathematics **58**, (New York: Springer–Verlag, 1977) and *p-adic Analysis: a Short Course on Recent Work*, London Mathematical Society Lecture Note Series **46** (Cambridge: Cambridge University Press, 1980) both by Neal Koblitz, require only a little more mathematical maturity.

VIII.14 Mahler is responsible for the criterion as to when a function can be interpolated to yield a p-adic analytic function. The question is this. Suppose that we have a sequence (f_n) of rational numbers. Is there a power series $f(t) = \sum_{k\geq 0}\phi_k t^k$, with positive radius of p-adic convergence so that $f(n) = f_n$? The argument is cute and, given its importance, surprisingly simple. I'll need the forward difference operator Δ, which acts on a sequence by $\Delta : f(n) \mapsto f(n+1) - f(n)$. We met it already in Lecture IV when I was chatting about the Bernoulli numbers. The main thought we'll use is the root test for convergence of a power series, whereby our series $\sum_{k\geq 0}\phi_k t^k$ has radius of convergence ρ^{-1} when $\limsup \sqrt[n]{\phi_n} = \rho$.

With ord meaning ord_p, it follows that the series converges p-adically for all t with $\operatorname{ord} t > -r$ exactly when $\liminf n^{-1} \operatorname{ord} \phi_n = r$. But if

$$f(t) = \sum_{k\geq 0}\phi_k t^k, \quad \text{then} \quad \Delta^n f(t) = \sum_{k\geq n}\phi_k \Delta^n t^k .$$

We note that $\Delta^n t^k$ is a polynomial in t of degree $k - n$ and with every coefficient divisible by $n!$. It follows that necessarily

$$\liminf n^{-1} \operatorname{ord} \Delta^n f(\bar{0}) = \liminf n^{-1} \operatorname{ord} \phi_n + n^{-1} \operatorname{ord} n!$$

and we learn that the series converges for $\operatorname{ord} t > -r$ if and only if

$$\liminf n^{-1} \operatorname{ord} \Delta^n f(\underline{0}) = r + 1/(p-1) .$$

The _ is there to say that 0 is the thing Δ'd. The argument just sketched provides a criterion for p-adic analyticity of the function $f(t)$ in terms of the values f takes at nonnegative integers, and thus a criterion for when these values can be interpolated to yield a p-adic analytic function.

---

*The primary source is currently `ftp://megrez.math.u-bordeaux.fr/pub/pari` and, for the Macintosh version, `ftp://hensel.mathp6.jussieu.fr/dist/pari` .

VIII.15 You may feel surprised at my blandly saying something about $\operatorname{ord} n!$. But it's fairly clear that

$$\operatorname{ord} n! = \left\lfloor \frac{n}{p} \right\rfloor + \left\lfloor \frac{n}{p^2} \right\rfloor + \left\lfloor \frac{n}{p^3} \right\rfloor + \left\lfloor \frac{n}{p^4} \right\rfloor + \cdots .$$

One now needs to watch carefully as one gets rid of those integer parts to conclude that

$$\operatorname{ord} n! = \frac{n - s(n)}{p - 1}$$

with $s(n)$ the sum of the digits of n presented in base p. In detail, write $n = \sum n_i p^i$ (of course, after a while all the n_i are zero). Now notice that

$$\frac{n}{p^j} - \left\lfloor \frac{n}{p^j} \right\rfloor = \frac{n_0}{p^j} + \frac{n_1}{p^{j-1}} + \cdots + \frac{n_{j-1}}{p} .$$

VIII.16 Since I've said so much already, let me add a few remarks on the p-adic exponential function and logarithm. The exponential function $\exp t$ is of course defined as usual, by the power series $\sum_{n=0}^{\infty} t^n/n!$. Now, it is a wonderful and easy-to-prove fact that in a nonarchimedean metric a series $\sum u_n$ converges if and only if $|u_n| \to 0$; or what is the same thing, if and only if $\operatorname{ord} u_n \to \infty$. In the p-adic cases there is no nonsense such as in the real case, where the series $\sum 1/n$ diverges notwithstanding that its terms go to 0. So $\exp t$ is defined if and only if $\operatorname{ord} t^n/n! \to \infty$. That is so if and only if $\operatorname{ord} t > 1/(p-1)$. So the p-adic exponential does not exist for all t, a situation much less pleasant than the familiar case. The p-adic logarithm is defined by the series $-\log(1-t) = \sum_{n\geq 1} t^n/n$ and plainly converges for $\operatorname{ord} t > 0$. Since $\operatorname{ord}\log(1-t) = \operatorname{ord} t$ we have $\exp\log(1-t) = 1-t$, or $\log(\exp t) = t$, only if $\operatorname{ord} t > 1/(p-1)$.

VIII.17 Kurt Mahler came to Australia in the early sixties, when I was an undergraduate. I felt lucky to be sent to see him at the Australian National University during a year that my Ph.D. supervisor was on sabbatical leave; doubly lucky, actually, as there was a friend I rather wanted to see in Canberra at the time. Mahler and I got along quite well, not so much because of our mathematics, but principally I fear, because of a common interest in science fiction. The appendix to my obituary 'Kurt Mahler 1903–1988'[*J. Austral. Math. Soc.* **51** (1991), pp. 343–380], is Mahler's autobiographical remarks '50 years as a mathematician, II'; it makes interesting reading.

VIII.18 By a helpful coincidence, soon after writing this lecture I received the July, 1993 copy of the *Bulletin of the American Mathematical Society* — the tyranny of distance dictates a 3-4 month delay — and there at pp. 14–50 saw Barry Mazur's survey 'On the passage from local to global in number theory'. One turns to that survey to see just how the local-global notions vaguely alluded to here are so important.

Notes on Fermat's Last Theorem

LECTURE IX

My brother Esau is a hairy man, but I am a smooth *man.*
from *Beyond the Fringe*

One of the difficulties in reading, or listening to, mature mathematics is its immense vocabulary and the volume of notions that seem to be required. Nor can one readily discover the meaning of the more popular ideas; all too often they are defined in terms of yet more obscure words. The truth is, fortunately, that few — perhaps none — of us know *all* the definitions. We rely on a feeling for what must be intended, knowing that we can refine that feeling should needs be. In a sense, these notes should be seen as little more than an attempt to create some useful feelings.

For example, it takes a while to shake off the belief that the calculus is only a matter of lots of tricks and many formulas. In fact, calculus is the acknowledgment that anything but a linear function is far too complicated for us to handle. Differentiation provides the techniques to tame functions by making them locally linear, and integration comprises the rules for sticking the local pieces together again.

This viewpoint also has the merit of suggesting that the purpose of linear algebra is not just to make it more difficult to solve systems of linear equations. At its most primitive, we have error estimation as in $f(x) - f(x_0) \simeq f'(x_0)(x - x_0)$. Change of variable x to $u = u(x)$ in an integral, and writing $du = u'(x)dx$ is a "sophisticated" version of the same thing. These matters come to proper flower in the case of several variables when one learns that the vector **du** is obtained from the vector **dx** by the action of the Jacobian matrix. So calculus is good for telling us that locally at $P_0(x_0, y_0, z_0)$ the good function $f(x, y, z) - f(x_0, y_0, z_0)$ behaves as does

$$\frac{\partial f}{\partial x}(x - x_0) + \frac{\partial f}{\partial y}(y - y_0) + \frac{\partial f}{\partial z}(z - z_0)\,.$$

But it's severe bad luck if *all* three partial derivatives vanish at P_0. Then our remark would have f pretend to be locally constant at P_0, which just ain't so for a decent f unless f were globally constant to begin with.

Alternatively, one admits to looking at the Taylor expansion of f at P_0 and one begins to consider higher-degree terms. That's too hard. So we declare P_0 a *singular* point of the function f and we resolve, as far as is practicable, to consider just *smooth* functions, namely those without singularities.

I want to say a few things about *curves*, that is polynomial equations $f(x, y) = 0$. The idea is to give some slight hint why smooth cubic curves with a rational point — elliptic curves — are very special indeed. By the way, I will suppose throughout that $f(x, y)$ is irreducible since otherwise our equation effectively represents more than one curve. Because we're interested in diophantine questions we also suppose that f is defined over $\mathbb{Q}$; hence after clearing denominators, it will be defined over $\mathbb{Z}$.

Curves have a *degree*, namely the total degree of the polynomial in its variables. That allows me to announce Bezout's theorem, according to which curves of degree m and n, respectively, intersect in at most mn points — indeed, they intersect in exactly mn points if we define and count points appropriately.

By "defining" points I mean two things. We know full well, for example, that a line (a curve of degree 1) may well seem not to intersect a circle (curve of degree 2) at all. But it should, twice. The answer is that we have failed to include complex points. Actually, given that our curves are defined over $\mathbb{Z}$, it would certainly suffice to allow points to have coordinates in the field $\overline{\mathbb{Q}}$ of all algebraic numbers. In the event, we won't want to do this, because below we'll only be interested in rational points; bad luck for Bezout. Second, we might realize that two lines, because both are of degree 1, ought in all decency to intersect in exactly one rational point. But they may well be parallel. We'll deal with that irritation by always being prepared to replace x by X/Z and y by Y/Z, and for a curve of degree n to multiply by Z^n. That turns $f(x, y)$ into a *homogeneous* polynomial of degree n in X, Y, and Z. Our parallel lines now intersect perfectly happily at a point. Of course that point has Z coordinate $Z = 0$; that is, it is on the *line at infinity*. One may refer to the new coordinates $(X : Y : Z)$ as *projective* coordinates. For $d \neq 0$ one identifies points $(a : b : c)$ and $(da : db : dc)$; now $(0 : 0 : 0)$ is not a point at all. Actually, that view seems old-fashioned. It's perfectly all right to drop the capitalization and just to deal with homogeneous polynomials $f(x, y, z)$ in the first place. Now, of course, a trivial solution $(0, 0, 0)$ really is irrelevant, because it did not arise from a point at all.

To an amateur's eye there might seem to be a special case in which a line intersects a circle just once. Of course, we are sophisticated enough to realize that the line is then tangent to the circle and that the point of intersection is to be counted twice. In that spirit, too, intersection through points where one or both of the curves is singular is to be counted to appropriate multiplicity.

Now I can turn to some elementary classical observations. The first is that the general curve of degree n has $\frac{1}{2}(n + 1)(n + 2)$ terms. Hence we should expect to be able to determine a curve of degree n by being given one fewer point through which it passes; that is, $\frac{1}{2}n(n+3)$ linear conditions on its coefficients should suffice. But even this is not always so. In the special case $n = 3$ we might then believe that nine points suffice, yet by Bezout two cubic curves C and C' intersect in precisely nine points. Hence such nine points of intersection determine the entire *pencil* $\mu C + \lambda C'$ of linear combinations of the two intersecting curves. This numerical accident might be thought of as being responsible for the richness of the structure of cubic curves.

But it does turn out that the nine points of intersection of a pair of cubic curves have a tight structure, in that any third cubic curve passing through eight of the points is necessarily an element of the said pencil of curves and thus also passes through the ninth point. From this follows the associativity of addition on cubic curves, whence the group structure on the set of rational points on such curves. The idea is to consider the triples of lines (l_1, l_2, l_3) and (m_1, m_2, m_3) cutting a cubic curve E as in the following tabulation:

A	B	$-A-B$	l_1	B	C	$-B-C$	m_1
$-A-B$	0	$A+B$	m_2	$-B-C$	0	$B+C$	l_2
$A+B$	C	$-(A+B)-C$	l_3	$B+C$	A	$-A-(B+C)$	m_3

Notice that I have defined addition of points by the rule $P + Q + R = 0$ if there is a line cutting E at P, Q, and R. The nine points $L \cdot E$ and $M \cdot E$, in which cubics L and M containing the points of the lines l (respectively, m) intersect E, have eight points in common. Hence the ninth points of intersection must also coincide, so that we have $A + (B + C) = (A + B) + C$.

Next we can remark that a curve of degree n has at most $\frac{1}{2}(n-1)(n-2)$ singular points. The idea is to intersect the given curve with an *adjoint* curve — one that passes through all the singular points of the given curve. Consider an adjoint of degree $n - 2$. Being of degree $n - 2$, it certainly can tolerate $\frac{1}{2}(n-2)(n+1)$ conditions. Were they available, it could thus be made to pass through as many as m singular points and an additional $\frac{1}{2}(n-2)(n+1) - m$ ordinary points of the given curve. But since the singular points are at the very least double points, it then intersects the given curve in at least $2m + \frac{1}{2}(n-2)(n+1) - m$ points counted according to multiplicity. By Bezout this number is at most $n(n-2)$, so we conclude that $m \le \frac{1}{2}(n-1)(n-2)$, as alleged.

One counts singular points appropriately. Ordinary points are not counted at all, and we say that a singularity is of order s if all partial derivatives of order $s - 1$ or less vanish, but there is a nonzero partial derivative of order s. A singularity of order s is counted $s - 1$ times.

But that's only the beginning of the story. The sad fact is that even when a singularity is just a *double point*, $s = 2$, the curve may well pass through that point more than once. One actually has to look closely to *resolve* the singularity. One fairly highbrow trick is to give the curve more freedom by thinking of it as living in additional dimensions. That gives one room to separate pieces of the curve that had come into inadvertent contact with one another. In summary, all curves have a smooth model whose projection back into two dimensions gives the original curve.

Alternatively, there is the method of local coordinates. In effect, one changes coordinates to bring the singularity to the origin and one then compares the situation with a known list. Finally, one prepares the list by looking closely.

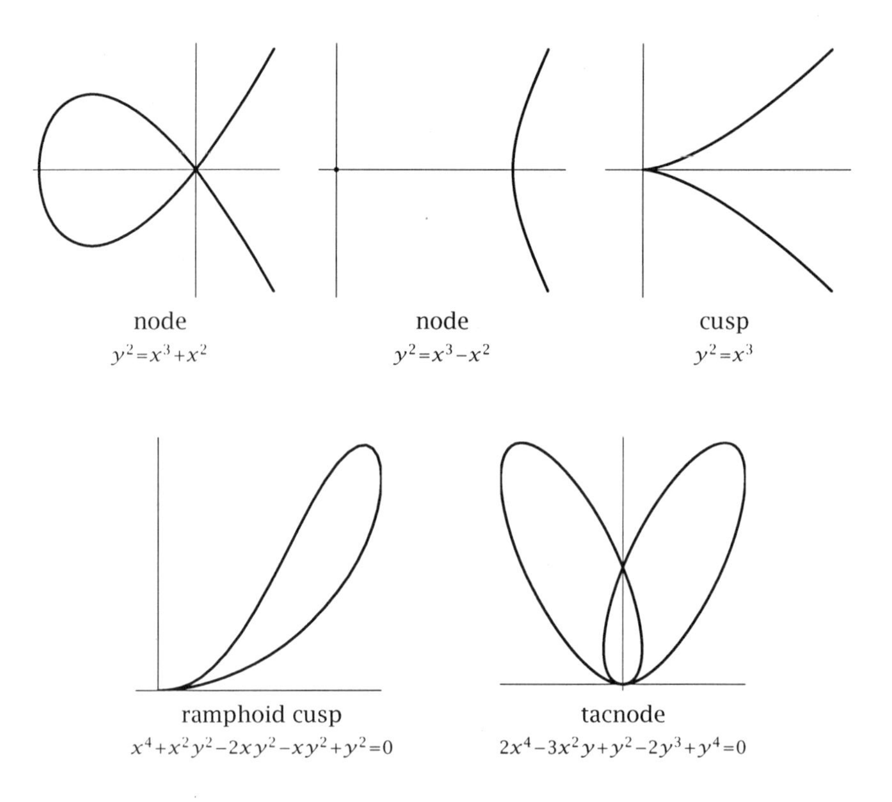

A variety of double points*

Of course, even supposing the only singularities are double points, a curve might not have as many double points as the number $\frac{1}{2}(n-1)(n-2)$ it is permitted. The defect, $\frac{1}{2}(n-1)(n-2)$ minus the actual number m of double points, is called the *genus* of the curve. For example, a cubic curve can tolerate one double point, so if it is smooth, then it has genus 1.

*Suggested by Robert J. Walker, *Algebraic Curves* (Princeton, N.J.: Princeton University Press, 1950).

But we mustn't forget that we are primarily interested in the structure of the set of rational points on our curve. So suppose that we stumble upon triples of rational functions

$$X = a(x,y,z), \quad Y = b(x,y,z), \quad Z = c(x,y,z),$$

and

$$x = A(X,Y,Z), \quad y = B(X,Y,Z), \quad z = C(X,Y,Z).$$

Then the curves $f(x,y,z)$ and $F(X,Y,Z)$ obtained one from the other by this *birational* transformation are said to be *equivalent*. Indeed, a birational transformation provides a $1:1$ correspondence between the points of f and F, other than for the singular points. In particular, it happens that one can always find a birational transformation to resolve all singularities into just double points. More important — this is of the essence — the genus of a curve turns out to be *invariant* under birational transformation.

One can now demonstrate various fundamental facts. For example, it is a theorem of Hurwitz and Hilbert that every curve of genus 0 is equivalent to a line or a conic. The trick is to show that such a curve of degree n is birationally equivalent to a curve of degree $(n-2)$.

In this spirit one shows that every rational curve C, of genus 1 and with a rational point S, is equivalent to a smooth cubic curve. Suppose that the curve C is of degree n. An adjoint of degree $(n-2)$ with triple intersection at S meets C in a further $(n-3)$ points. Now consider adjoints of degree $(n-2)$ through those $(n-3)$ points. This is a two parameter family, so of the shape $a_0\phi_0 + a_1\phi_1 + a_2\phi_2$. The family intersects C in three points. Then $\xi_0 : \xi_1 : \xi_2 = \phi_0 : \phi_1 : \phi_2$ is a birational transformation which changes C into a cubic curve.

Let me make these remarks explicit for the curve $C : y^2 = 2x^4 - 1$ which I was too lazy to cope with in a previous lecture.* Its only singularity, at $(0:1:0)$, turns out to be a pair of double points, so that C has genus 1. At the rational point $\mathcal{O}(1,1)$ on C we have $dy/dx = 4$ and $d^2y/dx^2 = -4$. By reason of its degrees in x and y, then $C' : y - 1 = -2(x-1)^2 + 4(x-1)$ is an adjoint, here displayed so that we can readily confirm it to have triple intersection with C at $\mathcal{O}$. The intersection at the double points counts as four common points, so there is one more point of intersection. That turns out to be the necessarily rational point $T(13, -239)$.

We can now see the group law on C. Each adjoint to C and passing through T cuts C at three further points P, Q, and R. The group law is given by $P + Q + R = \mathcal{O}$. To straighten matters out, we will move $\mathcal{O}$ to be a point at infinity. We first rewrite C' by setting

$$Z = (y + 239) + 2(x - 13)^2 - 44(x - 13).$$

---

*But I was clever enough to challenge my colleague Rod Yager to deal with the matter. Within moments, barely a fortnight's work and thought, it was done.

Now $Z = 0$ is an adjoint to C through T, having triple intersection at $\mathcal{O}$. Next, if $Y = 24(y + 239) + 32(x - 13)^2 + 856(x - 13)$, then $Y = 0$ is an adjoint to C through T, chosen so that it passes through each of the three *2-division points* of C; that is, the three points D satisfying $2D = \mathcal{O}$. $X = 8(y+239)+12(x-13)^2+304(x-13)$ is an adjoint through T chosen to give a congenial final result. In particular, it happens that the 2-division points of C are rational. So among other things we do arrange that $X = 0$ passes through one of these points $(-1, 1)$. Finally, one induces one's computer to compute x and y as rational functions of X, Y, and Z. After substitution and clearing denominators, miraculously one beholds that $Y^2Z = X^3 + 8XZ^2$.

Whatever, we can now show that indeed every curve of genus 1 with a rational point is birationally equivalent to a curve $y^2 = x^3 + ax + b$ in *Weierstrass form*. But such a cubic curve may be parametrized by elliptic functions $y = 4\wp'(z; \tau)$ and $x = 4\wp(z; \tau)$.

Let me confess to slurring an important issue. At the end of it all we're interested in diophantine questions, so we are concerned at being able to carry out the various tricks described while retaining rationality of the coefficients involved. I didn't want to pepper my remarks with the qualifier "rational", so that aspect is not emphasized except when I insist that my curve of genus 1 have a *rational* point. Otherwise, of course, any old point would do.†

Notes and Remarks

IX.1 It is well known but readily forgotten that the number of ways of choosing r things from n objects, with repetition permitted, is $\binom{n+r-1}{r}$. The traditional way to "see" this is to note that any selection a_0, a_1, a_2, $\ldots$, a_{r-1} from the symbols 1, 2, $\ldots$, n may be supposed nondecreasing. But such a selection corresponds $1 : 1$ with a selection $a_0 + 0, a_1 + 1, \ldots, a_{r-1} + (r - 1)$ of r distinct things now chosen from $n + r - 1$ objects. Hence the number of terms of degree n in three variables is

$$\binom{3+n-1}{n} = \tfrac{1}{2}(n+1)(n+2).$$

IX.2 To show that every curve of genus 0 is equivalent to a line or a conic, one shows that each such curve of degree n is birationally equivalent to a curve of degree $(n - 2)$.

---

† Originally I finished this lecture, right smack at the end of page four, with the remark: "But I see that to give a smooth rendition I've nearly gone into higher dimension myself, or at least beyond my page limit." It's too bad that this is no longer applicable.

One first notes that an adjoint of degree $(n-2)$ passes through an additional $(n-2)$ points, which we may nominate, on the curve. That provides a family $a_0\phi_0 + a_1\phi_1 + \cdots + a_{n-2}\phi_{n-2} = 0$ of adjoint curves of degree $(n-2)$, where the a_i are arbitrary constants and the ϕ_i are polynomials of degree $(n-2)$; we can guarantee the linear independence and irreducibility of the ϕ_i by our choices of the additional points. But a curve in the homogeneous coordinates $(\xi_0, \xi_1, \cdots, \xi_{n-2})$ defined by the $(n-2)$ equations $(\xi_0 : \xi_1 : \cdots : \xi_{n-2}) = (\phi_0 : \phi_1 : \cdots : \phi_{n-2})$ is of degree $(n-2)$. To see that, notice that the curve meets an arbitrary hyperplane $\sum a_i\xi_i = 0$ in points determined by the intersection of $\sum a_i\phi_i = 0$ and the original curve. Of the $n(n-2)$ points of intersection, $(n-1)(n-2)$ are given by the double points leaving $(n-2)$ variable intersections depending on the a_i. So taking any two of the equations

$$\frac{\xi_0}{\phi_0} = \frac{\xi_1}{\phi_1} = \frac{\xi_2}{\phi_2},$$

eliminating x, y, and z yields a curve $g(\xi_0, \xi_1, \xi_2) = 0$ of degree $(n-2)$, with the coordinates (ξ_0, ξ_1, ξ_2) depending birationally on (x, y, z).

IX.3 Checking on some of the ideas mentioned in this Lecture put me into a time warp. I was taught algebraic geometry — certainly I cannot say with any honesty that I learned algebraic geometry — from a text *Algebraic Curves*, by J. G. Semple and G. T. Kneebone (Oxford: Oxford University Press, 1959), some 30 years ago. Reopening that book, I read that "if a cubic curve C has an inflexion I then an allowable coordinate system can be chosen such that the equation of C takes the form

$$y^2z = m(x-\alpha z)(x-\beta z)(x-\gamma z).$$

The reference point is at I, YX is the inflexional tangent and ZX is the harmonic polar of I." By now, this is all algebra to me.* The amazing fact is that I got quite a good grade; but like Ozymandius, little remains.

Suppose that $f(x, y, z) = 0$ defines a curve C of degree n. The story seems to be that if $a = (a_x, a_y, a_z)$ is any point in the plane then the equation

$$\Big(a_x\frac{\partial}{\partial x} + a_y\frac{\partial}{\partial y} + a_z\frac{\partial}{\partial z}\Big)^r f(x, y, z) = 0,$$

if not satisfied identically, defines "a curve $C^{(r)}(a)$ called the rth polar of a with respect to C, or the polar $(n-r)$-ic of a ... ." It is fun that the equation of this polar may also be written as

$$\Big(x\frac{\partial}{\partial a_x} + y\frac{\partial}{\partial a_y} + z\frac{\partial}{\partial a_z}\Big)^{n-r} f(a_x, a_y, a_z) = 0.$$

It follows that the (first) polar of a cubic curve is a conic. One now notices that a nonsingular point P of a curve C is an inflection exactly if the polar

*Felicitously, this is a Dutch way of saying that it is all Greek to me. That is, I don't understand a word of it.

conic at P splits into a pair of lines, one of which is the tangent at P; the other line is called the *copolar line* of P. With that, the translation of the remark quoted above seems to amount to the fact that if a cubic is in Weierstrass form, then the flex tangent at infinity is the line $z = 0$ and the line $y = 0$ through the three 2-division points turns out to be its copolar.

IX.4 Let's be quite concrete for a moment and consider the graph of the curve $y^2 = f(x) = x^3 + Ax + B$. By "graph" one means the set of all points $(x, y) \in \mathbb{R}^2$, so of real points on the curve. Denote the discriminant $-16(4A^3+27B^2)$ by Δ. If $\Delta < 0$, then f has just one real zero and the graph has just one component. If $\Delta > 0$, then f has three distinct real zeros and the graph has two connected components. One is oval shaped and the other contains the point O at infinity. The three zeros of f are exactly the points of order 2 on the curve (and the inflection points correspond to the points of order 3).

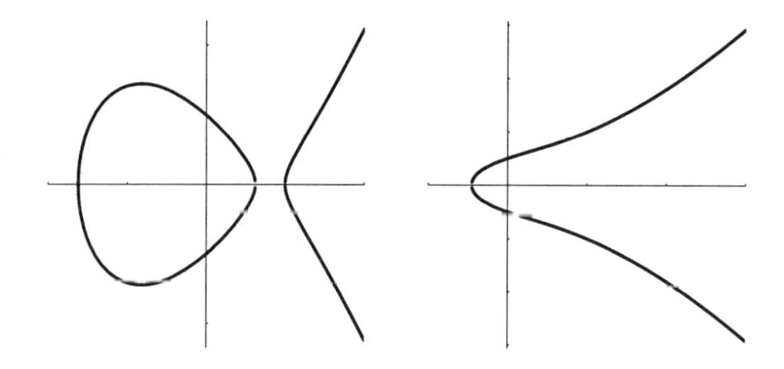

Possibilities for the real locus of an elliptic curve

Finally, if $\Delta = 0$, then we no longer have an elliptic curve, since the cubic curve now has a singular point. In this case

$$f(x) = (x - \alpha)^2(x - \beta)$$

and there are three cases. If $\alpha > \beta$, the curve has just one connected component with a double point at $x = \alpha$. The tangents at the double point have distinct real slopes. If $\alpha < \beta$, the curve has two connected components, with one consisting just of the point $(\alpha, 0)$; it is a double point with distinct complex tangents. Finally, if $\alpha = \beta$, then $2\alpha + \beta = 0$ entails that $\alpha = \beta = 0$ and the curve has a *cusp* at $(0, 0)$. That is, the tangents at the singular point coincide.

The three cases (with $\alpha = 0$) are the first three of "a variety of double points" displayed above, in this Lecture.

IX.5 The rational points on a cubic curve always have a group structure. If the curve is singular, so of genus 0, there are two cases for the group. In the case when f has two distinct zeros the group is isomorphic to the multiplicative group $\mathbb{Q}^\times$ of nonzero rationals; that is, the rational points may be parametrized by $\mathbb{Q}^\times$ in such a way as to preserve the group structure. If all three zeros of f coincide, then the group of rational points is isomorphic to the additive group $\mathbb{Q}^+$ of rationals.

IX.6 But why should we be interested in *singular* cubic curves? The point is that we will want to consider curves $y^2 = f(x) = x^3 + Ax + B$ originally defined over $\mathbb{Z}$ as if defined over finite fields $\mathbb{F}_p$; that is, we will consider *reductions* of the curve modulo p. If $\Delta \equiv 0 \pmod{p}$, the reduction will be singular and it is relevant to consider just how bad the reduction is. Disregarding $p = 2$ or 3, which are special (essentially because they're as small as the degree of the equation), there are two cases for the bad p. Either $f(x) \pmod{p}$ has two different zeros, the case of *multiplicative* reduction; or its three zeros coincide, the case of *additive* reduction. It's convenient to collect the bad primes and to combine the information as the *algebraic conductor* N of the curve by setting

$$N = \prod_{p \text{ bad}} p^{\gamma_p}.$$

If $p \neq 2$, 3, then the rule is that one takes $\gamma_p = 1$ for multiplicative reduction and $\gamma_p = 2$ in the case of additive reduction at p.

The algebraic conductor N of the curve E may be viewed as a convenient encoding of the places where E has bad reduction. It also encodes the nature of that reduction. If N is composed of distinct primes, so if E has multiplicative bad reduction at worst at any place, then the elliptic curve E is said to be *semistable*. Wiles' theorem asserts the truth of the Modularity Conjecture for *all* semistable elliptic curves defined over $\mathbb{Q}$. Because of a theorem of Ribet this entails Fermat's Last Theorem.

IX.7 Mind you, apart from my not saying anything about reduction at 2 and 3, my description cannot be quite right. Pretty obviously a given elliptic curve E is represented by many different equations $y^2 = f(x)$. For example, multiply by u^6 and set $y' = u^3y$, $x' = u^2x$. That yields a different model for E. At first glance (because of my plainly defective definition) E will now seem to have bad reduction at primes dividing u. Apparently, we will need some sort of notion of a *minimal model* for E, with primes of bad reduction for *that* model being the bad primes of E.

IX.8 In fact, the general cubic equation over any field is not $y^2 = f(x)$, but has the form

$$y^2 + a_1xy + a_3y = x^3 + a_2x^2 + a_4x + a_6.$$

One may refer to this as the "long" Weierstrass equation of the curve. It is now standard to set

$$\begin{aligned} b_2 &= a_1^2 + 4a_2\,, & b_4 &= 2a_4 + a_1a_3\,,\\ b_6 &= a_3^2 + 4a_6\,, & b_8 &= a_1^2a_6 + 4a_2a_6 - a_1a_3a_4 + a_2a_3^2 - a_4^2\,,\\ c_4 &= b_2^2 - 24b_4\,, & c_6 &= -b_2^3 + 36b_2b_4 - 216b_6\,. \end{aligned}$$

If the characteristic of the ground field (the field of definition) is not 2, we complete the square replacing $y + \frac{1}{2}(a_1x + a_3)$ by $\frac{1}{2}y$, to obtain

$$y^2 = 4x^3 + b_2x^2 + 2b_4x + b_6\,.$$

If the characteristic of the ground field isn't 3, then, on replacing (x, y) by $\left(\frac{1}{36}(x - 3b_2)\,, \frac{1}{108}\,y\right)$, we get

$$y^2 = x^3 - 27c_4\,x - 54c_6\,.$$

The *discriminant* Δ of the curve is defined by $1728\,\Delta = c_4^3 - c_6^2$, that is,

$$\Delta = -b_2^2b_8 - 8b_4^3 - 27b_6^2 + 9b_2b_4b_6\,.$$

Finally, the j-invariant of the curve is defined by $j = c_4^3/\Delta$.

IX.9 The original Weierstrass equation transforms into one of the same form by an *admissible* change of variables

$$x = u^2x' + r \qquad \text{and} \qquad y = u^3y' + s\,u^2x' + t\,.$$

Here u, r, s, and t belong to the ground field. As its name might suggest, the j-invariant is indeed invariant under these transformations.

Let $\mathbb{Q}$ be the *ground* field, that is, the field of definition. The problem now is to arrange for the Weierstrass model to have integral coefficients while achieving that $\operatorname{ord}_p \Delta$ is minimal for every p. This can be done one prime at a time; now *because* $\mathbb{Z}$ has unique factorization, these local minimizations do not interfere with one another. Thus one then obtains a global minimal model over $\mathbb{Z}$. This is the equation to which we refer when discussing the reduction at p of a given elliptic curve. Mind you, for p different from 2 or 3, no harm is done in working with equations of the shape $y^2 = x^3 - 27c_4x - 54c_6$.

For elliptic curves over arbitrary algebraic number fields there may be no globally minimal model and one must work with appropriate sets of locally minimal equations.

IX.10 Even though an elliptic curve may be defined over the rationals $\mathbb{Q}$, we can certainly pretend, if we find it convenient, that it is defined over some larger number field. If so, there are more admissible changes of variables and, possibly, fewer bad primes. A minimal model over $\mathbb{Q}$ may no longer be minimal over an extension field. In particular, there always is an extension field so that the curve does not have additive reduction anywhere. That's easy to see. It will do to think of the elliptic curve as being given by a short Weierstrass equation $y^2 = f(x)$. Additive reduction at p means that all three zeros of f coincide modulo p, so that, after a translation, they all are zero modulo p. Now replace x by px and y by $p^{3/2}y$, an admissible transformation if the ground field includes $\mathbb{Q}(\sqrt{p})$, and divide by p^3, thereby converting the additive reduction to multiplicative reduction at worst. The same effect, incidentally, is achieved by staying over $\mathbb{Q}$ and studying the *twist* $py^2 = f(x)$ of the original curve.

In any case, we see that additive reduction is not stable under change of ground field. Multiplicative reduction may be unstable. A consequence is that multiplicative reduction and, by courtesy, also good reduction are known as *semistable* reduction.

IX.11 A standard example taken from Knapp's book *vide infra* is the Fermat curve $x^3 + y^3 = z^3$. An ingenious substitution provides that

$$a^3 + (b + c)^3 = b^3 .$$

Now setting $x = a/c$, $y = b/c$ first yields $3y^2 + 3y = -x^3 - 1$; then after rescaling, the Weierstrass minimal model becomes

$$y^3 - 9y = x^3 - 27 .$$

A model which is minimal other than at 2 and 3 is $y^2 = x^3 - 432$.

IX.12 Suppose that $a^p + b^p + c^p = 0$, with $abc \neq 0$ and p an odd prime at least 5. The Frey curve $y^2 = x(x - a^p)(x + b^p)$ has a minimal model

$$y^2 + xy = x^3 + \tfrac{1}{4}(b^p - a^p - 1)x^2 - \tfrac{1}{16}a^p b^p x .$$

It suffices to replace x by $4x$ and y by $4x + 8y$ and then to divide by 2^6. Mind you, we'll need some conditions. Since we may suppose that a, b, and c have no common factor, just one of a, b, or c is even; and we may suppose b even, so $16 \mid b^p$, and a odd. Next, either a or $-a$ is $-1 \pmod 4$ and by multiplying all of a, b, and c by -1 if necessary — which loses no generality since p is odd — we may suppose that $a^p \equiv -1 \pmod 4$.

The invariants associated with this equation are

$$\begin{aligned} c_4 &= -(a^p b^p + b^p c^p + c^p a^p) , \\ c_6 &= -\tfrac{1}{2}(a^p - b^p)(b^p - c^p)(c^p - a^p) , \\ \Delta &= 2^{-8} a^{2p} b^{2p} c^{2p} . \end{aligned}$$

It is not difficult to check that the Frey curve has multiplicative reduction at the odd primes dividing abc. At 2 the model reduces to $y^2 + xy = x^3$ if $a^p \equiv -1 \pmod 8$ and to $y^2 + xy = x^3 + x^2$ if $a^p \equiv 3 \pmod 8$. In either case there are distinct tangents at the singular point $(0, 0)$. So there is also multiplicative reduction at 2. Thus the Frey curve is semistable.

In any case, $\gcd(c_4, \Delta) = 1$ implies immediately that the model we have obtained is minimal and defines a semistable elliptic curve. Specifically, it has good reduction at primes l not dividing $\frac{1}{16}a^p b^p c^p$ and bad reduction of multiplicative type at the primes dividing $\frac{1}{16}a^p b^p c^p$. If we write $N(n)$ for the "radical" or "conductor" of n, the product of the different primes dividing n, then the Frey curve has conductor $N = N\left(\frac{1}{16}abc\right)$.

IX.13 The best method of research is to look it up. But where? Some weeks after working on the equation $y^2 = 2x^4 - 1$, ill-timed good fortune happened to find me in the library at the Mathematisches Forschungsinstitut, Oberwolfach at around the letter "K". There I noticed the book *Elliptic Curves* (Mathematical Notes **40** (Princeton, N.J.: Princeton University Press, 1992), by Anthony W. Knapp. At page 55, I read that the curve $v^2 = 2u^4 - 1$ becomes $y^2 = x^3 + 8x$ on setting

$$x = \frac{2(v + 2u^2 - 1)}{(u-1)^2}, \qquad y = \frac{4((2u-1)v + 2u^3 - 1)}{(u-1)^3}.$$

Conversely,

$$u = \frac{y - 2x - 8}{y - 4x + 8}, \qquad v = \frac{y^2 - 24x^2 + 48y - 16x - 64}{(y - 4x + 8)^2}.$$

Had I seen this a little earlier, Rod Yager would have saved a lot of time and both of us would have learned a great deal less.

IX.14 Birational transformations are rather worrying. Where do they come from? Let me try a mildly complicated explanation. Traditionally, given a curve one studies "rational functions on the curve". The idea is that any quotient $r(x,y)/s(x,y)$ of polynomials is a perfectly good function on the plane (even where it blows up one can declare it to take the fairly harmless value ∞). Being quite good on the plane, it's also quite good defined just on some subset of the plane, such as just on all (x,y) which satisfy some irreducible equation $f(x,y) = 0$, that is, on the given curve defined by f. Mind you, we had better insist that f does not divide the denominator $s(x,y)$. Now, of course, something strange happens. Rational functions that look quite different, indeed *are* different on the plane, may become the same when restricted to the curve. In fact, r/s and u/v define the same function on the curve exactly when f divides $ru - sv$. Given that f *is* irreducible — "prime" — the upshot is that the collection of rational functions on the curve forms a field.

Such a field of functions is an algebraic extension $\mathbb{Q}(x,y)$ of the field of rational functions $\mathbb{Q}(x)$ in one variable. Here x and y are distinguished as the generators of the field. But there are plenty of other choices $X(x,y)$ and $Y(x,y)$ of rational functions X and Y in (x,y) that serve just as well as generators. Now $\mathbb{Q}(x,y) = \mathbb{Q}(X,Y)$ entails also that $x = x(X,Y)$ and $y = y(X,Y)$ are rational functions in (X,Y). This is how birational transformations arise naturally. One simply chooses X, Y to be alternative generators of the field of functions on the curve.

IX.15 A possibly more instructive way of explaining associativity of the group law on curves of genus 1 consists of going back to very first principles. Given points A, B, and C on a nonsingular cubic, the line $l = \overline{AB}$ gives a third point of intersection, U, say. Now the line $\overline{CU}$ determines a fourth point, D. More abstractly, one can construct a function on the curve with poles at A and B and a zero at C. There must then be a further zero, indeed D. The fact that *one* point is determined is a manifestation of genus *one*. One now proceeds to define the group law $A + B = C$ by fixing a rational point O and letting A, B, and O determine C. That $D = A + (B + C) = (A + B) + C$ follows immediately by considering a function with poles at A, B, C and zeros at O, O, and a third point D.

Pierre de Fermat

detail from engraving by F. Poilly
Varia Opera Mathematica D. Petri de Fermat ... , Tolosæ 1679.

the equation $E(j')$ is still minimal and semistable at 5, since a criterion for this, for an integral model, is that either $\mathrm{ord}_5(\Delta(E(j'))) = 0$ or $\mathrm{ord}_5(c_4(E(j'))) = 0$. So up to a quadratic twist E' is also semistable. □

This kind of argument can be applied more generally.

THEOREM 5.3. *Suppose that E is an elliptic curve defined over $\mathbf{Q}$ with the following properties:*

(i) *E has good or multiplicative reduction at 3, 5,*

(ii) *For $p = 3,5$ and for any prime $q \equiv -1 \bmod p$ either $\bar{\rho}_{E,p}|_{D_q}$ is reducible over $\bar{\mathbf{F}}_p$ or $\bar{\rho}_{E,p}|_{I_q}$ is irreducible over $\bar{\mathbf{F}}_p$.*

Then E is modular.

Proof. The main point to be checked is that one can carry over condition (ii) to the new curve E'. For this we use that for any odd prime $p \neq q$,

$$\bar{\rho}_{E,p}|_{D_q} \text{ is absolutely irreducible and } \bar{\rho}_{E,p}|_{I_q} \text{ is absolutely reducible and } 3 \nmid \#\bar{\rho}_{E,p}(I_q)$$

$$\Updownarrow$$

$$E \text{ acquires good reduction over an abelian 2-power extension of } \mathbf{Q}_q^{\mathrm{unr}} \text{ but not over an abelian extension of } \mathbf{Q}_q.$$

Suppose then that $q \equiv -1(3)$ and that E' does not satisfy condition (ii) at q (for $p = 3$). Then we claim that also $3 \nmid \#\bar{\rho}_{E',3}(I_q)$. For otherwise $\bar{\rho}_{E',3}(I_q)$ has its normalizer in $\mathrm{GL}_2(\mathbf{F}_3)$ contained in a Borel, whence $\bar{\rho}_{E',3}(D_q)$ would be reducible which contradicts our hypothesis. So using the above equivalence we deduce, by passing via $\bar{\rho}_{E',5} \simeq \bar{\rho}_{E,5}$, that E also does not satisfy hypothesis (ii) at $p = 3$.

We also need to ensure that $\bar{\rho}_{E',3}$ is absolutely irreducible over $\mathbf{Q}(\sqrt{-3})$. This we can do by observing that the property that the image of $\bar{\rho}_{E',3}$ lies in the Sylow 2-subgroup of $\mathrm{GL}_2(\mathbf{F}_3)$ implies that E' is the image of a rational point on a certain irreducible covering of C of nontrivial degree. We can then argue in the same way we did in the previous theorem to eliminate the possibility that $\bar{\rho}_{E',3}$ was reducible, this time using two separate coverings to ensure that the image of $\bar{\rho}_{E',3}$ is neither reducible nor contained in a Sylow 2-subgroup.

Finally one also has to show that if both $\bar{\rho}_{E,5}$ is reducible and $\bar{\rho}_{E,3}$ is induced from a character of $\mathbf{Q}(\sqrt{-3})$ then E is modular. (The case where both were reducible has already been considered.) Taylor has pointed out that curves satisfying both these conditions are classified by the non-cuspidal rational points on a modular curve isomorphic to $X_0(45)/W_9$, and this is an elliptic curve isogenous to $X_0(15)$ with rank zero over $\mathbf{Q}$. The non-cuspidal rational points correspond to modular elliptic curves of conductor 338. □

A page from 'Modular elliptic curves and Fermat's Last Theorem'

Annals of Mathematics **141** (3), May 1995

LECTURE X

The Eisenstein series $G_{2k}(\tau)$ of Lecture VII have period 1 and so must have Fourier expansions. It's fun to discover that

$$G_{2k}(\tau) = \sum_{\omega\in\Omega}{}' \frac{1}{\omega^{2k}} = 2\zeta(2k) + 2\frac{(2i\pi)^{2k}}{(2k-1)!}\sum_{n=1}^{\infty} \sigma_{2k-1}(n)q^n .$$

Here $q = e^{2i\pi\tau}$ and $\sigma_r(n) = \sum_{d|n} d^r$ is the sum of the rth powers of the divisors of n. If we have the presence of mind to remember Euler's evaluation of $\zeta(2k)$, then we see, on normalizing to constant coefficient 1, that one has

$$E_{2k}(\tau) = -\frac{(2k)!}{B_{2k}(2i\pi)^{2k}}G_{2k}(\tau) = 1 - \frac{4k}{B_{2k}}\sum_{n=1}^{\infty} \sigma_{2k-1}(n)q^n .$$

We know the Bernoulli numbers $B_4 = -\frac{1}{30}$ and $B_6 = \frac{1}{42}$, so we can now compute

$$\begin{aligned}(2i\pi)^{12}&\Big((E_4(\tau))^3 - (E_6(\tau))^2\Big) / 12^3 \\ &= (60G_4(\tau))^3 - 27(140G_6(\tau))^2 = g_2^3 - 27g_3^2 .\end{aligned}$$

After normalizing by dividing by $(2i\pi)^{12}$, we obtain

$$\Delta(\tau) = (q - 24q^2 + 252q^3 - 1472q^4 + 4830q^5 - 6048q^6 - \cdots).$$

The "discriminant" Δ — it honest-to-goodness is the discriminant of the elliptic curve parametrized by $\wp(z;\tau)$ — is an example of a *cusp form*, a modular form (of weight 12 in this case) that vanishes at the "cusp" $\tau = i\infty$; that is, at $q = 0$. Incidentally, that Fourier series for Δ hides the beautiful product expansion

$$\Delta(\tau) = q\prod_{n=1}^{\infty}(1-q^n)^{24} .$$

Surely, there's magic afoot here.

Now notice that if $M\tau = (a\tau + b)/(c\tau + d)$, with M an integer matrix of determinant 1, so an element of $\mathrm{SL}(2,\mathbb{Z})$, then $\Delta(M\tau) = (c\tau + d)^{-12}\Delta(\tau)$. The multiplier $(c\tau + d)^{-12}$ is a sadness. But the modular form $(E_4(\tau))^3$

is also of weight 12 and produces the same factor. Hence the quotient $(E_4(\tau))^3/\Delta$, denoted by $j(\tau)$ and given by

$$j(\tau) = \frac{1}{q} + 744 + 196884q + 21493760q^2 + 864299970q^3 + \cdots ,$$

is *invariant* under the transformations M. It is a modular *function*.

The *automorphic* property, that is, the invariance of $j(\tau)$ under the transformations just described, generalizes periodicity. Just as we know an elliptic function once we have the values it takes on its period parallelogram, so we know the modular forms just described once we have their values on their fundamental domain. There is a nice side effect on the number of zeros and poles of these functions, in complete analogy with a property of rational functions on which I remarked in Lecture VIII. There I mentioned that for a rational function f — a quotient of polynomials — we have the formula $\sum_{\alpha \in \mathbb{C} \cup \infty} \operatorname{ord}_\alpha f = 0$.

In straightforward language this says that the number of zeros of f just equals its number of poles — because $\operatorname{ord}_\alpha f$ is negative at poles α. We had to "close" the complex plane by adding the point at ∞ to make this work out. There is a "moreover". Namely, a rational function in fact takes every value on the complex sphere the same number of times. That number — which we might call the *order* of the rational function — is, incidentally, the larger of the degrees of the numerator and denominator of f. In exactly this spirit, and in a sense by the same argument, the same kind of thing is true for elliptic and modular functions on their respective fundamental domains. A related cute fact true of elliptic functions is that if, in its fundamental parallelogram, its zeros are at points α_i and its poles at points β_i, then $\sum \alpha_i - \sum \beta_i$ is a period of the function, that is, a $\mathbb{Z}$-linear combination of τ and 1. On its period parallelogram, the Weierstrass $\wp$-function takes every value exactly twice. We can see that its order is precisely 2 because there is a pole of order 2 at the lattice points and the function is holomorphic elsewhere. Similarly, the derivative $\wp'$ has order 3.

A consequence of these observations is that an elliptic function

$$\wp'(z) - A\wp(z) - B$$

has precisely three zeros u, v, and w, and they satisfy $u + v + w \in \Lambda$. Hence if $u + v + w \equiv 0 \pmod{\Lambda}$, then

$$\begin{vmatrix} \wp(u) & \wp'(u) & 1 \\ \wp(v) & \wp'(v) & 1 \\ \wp(w) & \wp'(w) & 1 \end{vmatrix} = 0 .$$

Alternatively for either $z = u$, $z = v$, or $z = -u - v$ we have

$$\wp'^2(z) - (A\wp(z) + B) = 0 ,$$

so $4\wp^3(z) - A^2\wp^2(z) - (2AB + g_2)\wp(z) - (B + g_3) = 0$. Hence

$$\wp(u) + \wp(v) + \wp(u + v) = \tfrac{1}{4}A^2$$

and we have the *addition formula*

$$\wp(u + v) = \tfrac{1}{4}\left(\frac{\wp'(u) - \wp'(v)}{\wp(u) - \wp(v)}\right)^2 - \wp(u) - \wp(v).$$

In the special case $u = v$ we evidently obtain the *duplication* formula

$$\wp(2u) = \tfrac{1}{4}\left(\frac{\wp''(u)}{\wp'(u)}\right)^2 - 2\wp(u).$$

These formulas make explicit the addition rule on an elliptic curve.

It is in the spirit of the present observations to remark that every elliptic function $f(z;\tau)$ is of the shape $E(\wp(z;\tau)) + \wp'(z;\tau)F(\wp(z;\tau))$; here E and F are rational functions. In particular, for any integer n, $\wp(nz)$ is a rational function in $\wp(z)$. Analogously, every modular function is a rational function of $j(z)$.

On its fundamental domain, the modular invariant $j(z)$ has a pole of order 1 at infinity. We'll notice that it has a zero at $\rho = \frac{1}{2}(1 - i\sqrt{3})$ and takes the value 1 at i. But the main point is that each value in $\mathbb{C} \cup \infty$ is taken exactly once, so j provides a one-to-one map of its fundamental domain to the complex plane.

Thus, the situation is this. Suppose that we're handed a curve of genus 1. If there's no rational point on it, then we've already found all its rational points; so it's reasonable to suppose that we have a rational point O. Then the tricks of Lecture IX allow us to model the curve by a homogeneous equation $ZY^2 = X^3 + aXZ^2 + bZ^3$.

Formally, an *elliptic curve* defined over the rationals $\mathbb{Q}$ is precisely a curve of genus 1 with a rational point O. Equivalently, it is a curve with a group law or addition rule where O is the 0 of the group law. In the last lecture we saw that the points on a cubic curve do indeed have an addition rule and that given a point O on a curve of genus 1, we can find a birational transformation

$$(X : Y : Z) = (X(x, y, z) : Y(x, y, z) : Z(x, y, z))$$

so that the curve has an equation in Weierstrass form

$$ZY^2 = X^3 + aXZ^2 + bZ^3.$$

Given this form of the equation, the line $Z = 0$ has triple intersection with the curve at O. Thus its point O becomes a point of inflection, and the line at infinity the flex tangent. It also follows, as should already be clear, that the line $Y = 0$ is the line containing the three 2- *division points* — those points P satisfying $2P = 0$. In general, these points will not have rational coordinates.

For the curve in Weierstrass form we may set $Z = \frac{1}{4}$, $Y = \wp'(z;\tau)$, and $X = \wp(z;\tau)$ with τ given by $a = -4g_2 = -240G_4(\tau)$ and $b = -16g_3 = -2240G_6(\tau)$. One immediate worry might have been whether there is always a τ; and if there is, whether it is properly determined by knowing g_2 and g_3. However, all's well. Because j is one-to-one, there is a τ provided that the given curve is nonsingular, that is, if $4a^3 + 27b^2 \neq 0$ — I assumed this tacitly, for otherwise the equation $y^2 = x^3 + ax + b$ represents a curve of genus 0. Further, τ is as unique as it can be, which is up to transformations

$$\tau \mapsto \frac{a\tau + b}{c\tau + d} \quad \text{with} \quad \begin{pmatrix} a & b \\ c & d \end{pmatrix} \in \mathrm{PSL}(2,\mathbb{Z}).$$

By the way: tradition demands that the group of invertible $n \times n$ matrices with entries in $\mathbb{R}$, say, be called $\mathrm{GL}(n,\mathbb{R})$ or $\mathrm{GL}_n(\mathbb{R})$, the *general linear group* ... ; and that in that spirit one refers to the 2×2 matrices with determinant 1 and integer entries as the *special linear* group $\mathrm{SL}(2,\mathbb{Z})$. If we don't want to be bothered with both $\pm M$ — and we don't, because we're interested in the transformations, not the matrices — we have just the *projective linear* group. In a moment I'll forget all this and will refer to the group I've just called $\mathrm{PSL}(2,\mathbb{Z})$ as Γ, the full *modular* group.

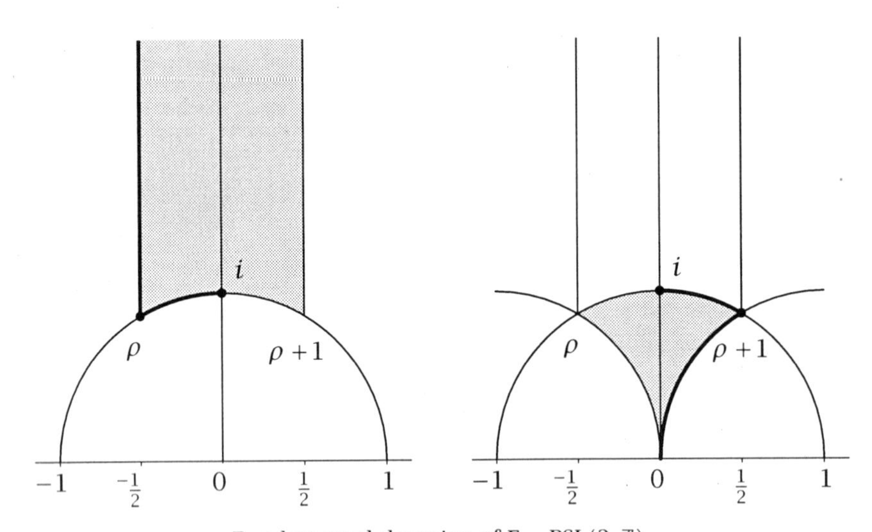

Fundamental domains of $\Gamma = \mathrm{PSL}(2,\mathbb{Z})$

Since a modular function has the same number of zeros as poles in its fundamental domain, it follows that modular forms f and g of weight k must have the same number of zeros in their fundamental domain. We need only notice that the quotient of two modular forms f, g of weight k is a modular function f/g. For example, Δ is of weight 12 for Γ and has just a simple zero, at infinity. Since the quotient Δ^k/f^{12} has a k-tuple zero at infinity it must be that f has $k/12$ zeros in its fundamental domain. It follows that, for a modular form f for Γ, we have

$$\tfrac{1}{12}k = \tfrac{1}{3}\operatorname{ord}_\rho f + \tfrac{1}{2}\operatorname{ord}_i f + \operatorname{ord}_{\tau\neq i,\rho} f\,.$$

The point is that, so to speak, just a third of the point ρ and only a half of the point i belongs to the fundamental domain $\mathcal{H}/\Gamma$.

It's obvious that this equation somewhat restricts the possibilities for the modular forms of weight k because zeros have to come in whole lots. Plainly, $k < 4$ is impossible and every modular form of weight 4 has a zero at $z = \rho$. So a quotient of any two modular forms of weight 4 has neither zeros nor poles and must therefore be a constant. Thus, up to normalization, G_4 is the only modular form for Γ of weight 4. Similarly, G_6 vanishes at $z = i$ and is the sole form of weight 6. Once again we see that, indeed, G_8 must be a constant multiple of G_4^2, and G_{10} a constant multiple of G_4G_6. For $k = 12$, on the other hand, there are the two possibilities G_4^3 and G_6^2, allowing us to form the linear combination Δ vanishing at infinity.

The modular forms for Γ of some given weight k form a vector space, denoted $M_k(\Gamma)$, whose dimension is $\lfloor k/12 \rfloor + 1$ if $k \not\equiv 2 \pmod{12}$ and is $\lfloor k/12 \rfloor$ otherwise. Morally, there ought to be a modular form of weight 2. Indeed, there is the entire function

$$G_2(\tau) = \sum_n{}' \sum_{m=-\infty}^{\infty} \frac{1}{(n\tau + m)^2} = 2\zeta(2) - 8\pi^2 \sum_{n=1}^{\infty} \sigma_1(n)q^n\,,$$

which makes sense if one sums the inner sum first. This is a case of a troublesome double sum. It nearly makes it, but not quite; it satisfies

$$G_2\Big(\frac{a\tau + b}{c\tau + d}\Big) = (c\tau + d)^2 G_2(\tau) - 2i\pi c(c\tau + d)\,.$$

Once the dimension of the space of modular forms is more than 1, so when $k \geq 12$ for modular forms with respect to Γ, we can arrange to concoct cusp forms. There is just the one vanishing condition, so the subspace $S_k(\Gamma)$ of cusp forms has dimension one less than that of the space $M_k(\Gamma)$.

We might well wonder why the fuss about cusp forms, that is, forms vanishing at the one cusp of Γ at infinity. A seemingly silly explanation might be to say that their Fourier expansions $\sum_{n=1}^{\infty} a_n q^n$ can be replaced, just like that, by a corresponding Dirichlet series

$$\sum_{n=1}^{\infty} \frac{a_n}{n^s}\,.$$

Mind you, given a cusp form $f(z) = \sum a_n q^n$, this "replacement" is actually effected by taking the *Mellin transform*

$$\int_0^{\infty} f(it)t^s \frac{dt}{t} = (2\pi)^{-s}\Gamma(s) \sum_{n=1}^{\infty} \frac{a_n}{n^s}\,.$$

It's now very relevant that $a_0 = 0$. Here $\Gamma(s)$, the Mellin transform of e^{-t}, is the classical Γ-function, analytically interpolating the factorial $(n-1)!$.

Notes and Remarks

X.1 First a reminder. While $G_{2k}(M\tau) = (c\tau + d)^{-2k}G_{2k}(\tau)$ produces that nasty factor, in the special case $c = 0$, $d = 1$ (so necessarily $a = 1$, but b arbitrary) the factor is just 1. So $G_{2k}(\tau + b) = G_{2k}(\tau)$ for all $b \in \mathbb{Z}$, which demonstrates the periodicity of the Eisenstein series. Hence, all being reasonably well, they have Fourier expansions. The sequence one follows in the argument is

$$\begin{aligned} G_{2k}(\tau) &= \sum_{(m,n)\neq(0,0)} \frac{1}{(n\tau + m)^{2k}} = \sum_{m\neq 0} \frac{1}{m^{2k}} + \sum_{n\neq 0} \sum_{m=-\infty}^{\infty} \frac{1}{(n\tau + m)^{2k}} \\ &= 2\zeta(2k) + 2\frac{(2i\pi)^{2k}}{(2k-1)!} \sum_{d=1}^{\infty} \sum_{a=1}^{\infty} d^{2k-1} q^{da} \\ &= 2\zeta(2k) + 2\frac{(2i\pi)^{2k}}{(2k-1)!} \sum_{n=1}^{\infty} \sigma_{2k-1}(n) q^n . \end{aligned}$$

The principal trick was to notice that

$$\pi \cot \pi\tau = i\pi \frac{q+1}{q-1} = i\pi - \frac{2i\pi}{1-q} = i\pi - 2i\pi \sum_{d=1}^{\infty} q^d$$

and then to remember the partial fraction expansion for $\pi \cot \pi\tau$, whence

$$\frac{1}{\tau} + \sum_{m=1}^{\infty} \left(\frac{1}{\tau + m} + \frac{1}{\tau - m} \right) = i\pi - 2i\pi \sum_{d=1}^{\infty} q^d .$$

One differentiates this identity with respect to τ to learn that

$$\sum_{m=-\infty}^{\infty} \frac{1}{(\tau + m)^{2k}} = \frac{(2i\pi)^{2k}}{(2k-1)!} \sum_{d=1}^{\infty} d^{2k-1} q^d .$$

X.2 The divisor functions $\sigma_r(n)$ are standard examples of *arithmetic* functions, properly beloved by amateurs though in this case appearing in a professional context. Indeed, arithmetic functions expect to appear as the coefficients of Dirichlet series. For example, it's not too hard to show — see the Notes following Lecture VIII — that one has

$$\sum \frac{\sigma_r(n)}{n^s} = \zeta(s - r)\zeta(s) .$$

Still, as I'll say several more times in the sequel, Dirichlet series are harder to handle than power series.

It's a joy to meet divisor functions as the coefficients of Fourier series, thus power series in q, representing modular forms. Thus identities such as $G_4^2 = \frac{7}{3}G_8$, which I originally mentioned for their amusement value, now become quite profound relationships connecting the divisor functions $\sigma_3(n)$ and $\sigma_7(n)$.

X.3 I cheated slightly in failing to emphasize that the Fourier coefficients of $\Delta(\tau)$, traditionally* denoted by $\tau(n)$ — the function τ of n so defined is known as the Ramanujan τ-function — really all are integers. Whether any is zero, and the matter of their rate of growth, were long subjects of conjecture. The latter question was settled by Deligne in the seventies. Integrality only requires checking somc fairly straightforward congruences. To see this, notice that explicitly

$$E_4(\tau) = 1+2^4\cdot 3\cdot 5\sum_{n\geq 1}\sigma_3(n)q^n \quad \text{and} \quad E_6(\tau) = 1-2^3\cdot 3^2\cdot 7\sum_{n\geq 1}\sigma_5(n)q^n .$$

Thus integrality is just a matter of being able to divide the appropriate combination by 12^3. With a teeny bit of care we see that this boils down to just whether

$$5\sigma_3(n) + 7\sigma_5(n) \equiv 0 \pmod{12} .$$

But

$$5\sigma_3(n) + 7\sigma_5(n) = \sum_{d\mid n}(5d^3 + 7d^5) \equiv \sum_{d\mid n} d^3(1-d^2) \pmod{12},$$

so since 12 always divides $d^3(1-d^2)$, we are done.

X.4 There's a beautiful essay of Jean-Pierre Serre on modular functions, Part III of his wonderful book†, *A Course in Arithmetic*, Graduate Texts in Mathematics **7** (New York: Springer–Verlag 1973). *Inter alia* Serre tells a story about $G_2(\tau)$. To start with, consider the double series

$$H_2(\tau) = \sum_n \sum_m{}' \frac{1}{(m-1+n\tau)(m+n\tau)},$$

$$H(\tau) = \sum_m \sum_n{}' \frac{1}{(m-1+n\tau)(m+n\tau)}$$

where the (m,n) run through all $m \in \mathbb{Z}$ and $n \in \mathbb{Z}$, with as usual the $\sum'$ indicating that silly terms, here $(m,n) = (0,0)$ and $(1,0)$, are simply to be omitted. Note the order of the summations. Because of

$$\frac{1}{(m-1+n\tau)(m+n\tau)} = \frac{1}{m-1+n\tau} - \frac{1}{m+n\tau}$$

one can discover using the expansions for $\pi\cot\pi\tau$ above that $H_2 = 2$ while $H = 2 - 2\pi i/\tau$. Naturally, this supposes τ in the upper half-plane.

*Though a little confusingly if τ also denotes our variable.

†Originally, *Cours d'arithmétique* (Paris: PUF 1970).

Now turn to the series

$$G_2(\tau) = \sum_n \sum_m{}' \frac{1}{(m+n\tau)^2} \quad \text{and} \quad G(\tau) = \sum_m \sum_n{}' \frac{1}{(m+n\tau)^2}.$$

Because a double series with general term

$$\frac{1}{(m-1+n\tau)(m+n\tau)} - \frac{1}{(m+n\tau)^2} = \frac{1}{(m-1+n\tau)(m+n\tau)^2}$$

is absolutely convergent, we must have $G_2 - H_2 = G - H$. So

$$G_2(\tau) - G(\tau) = H_2(\tau) - H(\tau) = 2\pi i/\tau.$$

Now $G_2(-1/\tau) = \tau^2 G(\tau)$ yields $G_2(-1/\tau) = \tau^2 G_2(\tau) - 2\pi i\tau$, which is what I wanted to show.

X.5 The functional equation for G_2 fairly readily yields also a functional equation for the product $F(\tau) = q\prod_{n=1}^{\infty}(1-q^n)^{24}$. It's wise to tame a product by taking a logarithm. On taking the derivative of that logarithm we obtain the logarithmic differential

$$\frac{dF}{F} = \frac{dq}{q}\Big(1 - 24\sum_{n=1}^{\infty}\sum_{m=1}^{\infty} nq^{nm}\Big) = \frac{dq}{q}\Big(1 - 24\sum_{n=1}^{\infty}\sigma_1(n)q^n\Big).$$

On the other hand,

$$G_2(\tau) = \frac{\pi^2}{3} - 8\pi^2\sum_{n=1}^{\infty}\sigma_1(n)q^n$$

so that

$$\frac{dF}{F} = \frac{6i}{\pi}G_2(\tau)d\tau.$$

Hence

$$\begin{aligned}\frac{dF(-1/\tau)}{F(-1/\tau)} &= \frac{6i}{\pi}G_2(-1/\tau)\frac{d\tau}{\tau^2}\\ &= \frac{6i}{\pi}(\tau^2 G_2(\tau) - 2\pi i\tau)\frac{d\tau}{\tau^2} = \frac{dF(\tau)}{F(\tau)} + 12\frac{d\tau}{\tau}.\end{aligned}$$

So the two functions $F(-1/\tau)$ and $\tau^{12}F(\tau)$ have the same logarithmic differential. Checking at $\tau = i$, noting that $F(i) \neq 0$, now confirms that $F(-1/\tau) = \tau^{12}F(\tau)$. Thus

$$q\prod_{n=1}^{\infty}(1-q^n)^{24}$$

is a cusp form of weight 12 for the full modular group Γ, and by uniqueness is indeed a product expression for the discriminant Δ.

X.6 I'll remark in Lecture XII that if $\alpha = \frac{1}{2}(\sqrt{-163}+1)$, then $j(\alpha)$ is a rational integer — because the domain $\mathbb{Z}[\alpha]$ has unique factorization. But the q-expansion for $j(\tau)$ shows that

$$j(\alpha) = -e^{\pi\sqrt{163}} + 744 + (\text{terms that are very small for } \tau = \alpha).$$

So $e^{\pi\sqrt{163}}$ should be very nearly an integer. Indeed,

$$e^{\pi\sqrt{163}} = 262\,537\,412\,640\,768\,743.999\,999\,999\,999\,2\ldots\,.$$

It is also fun to notice that

$$j(\alpha) = -262\,537\,412\,640\,768\,000 = (-640\,320)^3$$

is a perfect cube. For details, see H. M. Stark, 'Class number of complex quadratic fields', in *Modular Functions of One variable* I, (Proc. International Summer School, Antwerp 1972), Springer Lecture Notes in Mathematics **320** (New York: Springer-Verlag, 1973), pp. 153–174.*

X.7 Given a lattice Λ, so fixing τ, we consider all elliptic functions with that period lattice. It's easy to see that the quotient of two such functions and their difference yield meromorphic functions with period lattice Λ, so the set of elliptic functions forms a field. The remarkable fact is that every function f in this field is the sum of a rational function in $\wp$ and of $\wp'$ times some rational function in $\wp$. Showing this is quite simple. To begin with, $2g(z) = f(z)+f(-z)$ is an even function. Suppose that a is a zero and b is a pole of g. Then $g(z)(\wp(z)-\wp(b))/(\wp(z)-\wp(a))$ has the same zeros and poles as does $g(z)$, other than for its zeros equivalent to $\pm a$ and its poles equivalent to $\pm b$ with respect to the lattice Λ. Repeating this process leads after finitely many steps to an elliptic function without poles, that is, to a constant. So any even elliptic function is a rational function in $\wp$. Since $\wp'$ is an odd function, $2h(z) = (f(z)-f(-z))/\wp'(z)$ is an even function. Finally, $f = g + \wp' h$.

X.8 Similarly, given a function g modular with respect to the full modular group Γ, suppose that a is a zero and b is a pole of f. Much as above, it follows that $g(z)(j(z)-b)/(j(z)-a)$ has the same zeros and poles as does $g(z)$, other than for its zeros equivalent to a and its poles equivalent to b with respect to the transformation group Γ. That leads ultimately to a modular function on Γ without a zero or pole, so to a constant. Thus every modular function $g(z)$ on Γ is a rational function in $j(z)$.

X.9 Cubics have exactly nine flexes, with the interesting property that any line through two of them contains a third flex. It is a pleasant exercise to show that the flexes of any curve are given by the points of intersection of it with its *Hessian*, the curve

$$\frac{\partial^2 f}{\partial x^2}\frac{\partial^2 f}{\partial y^2} - \frac{\partial^2 f}{\partial x\,\partial y} = 0\,.$$

See my meanderings on polar curves in the Notes to Lecture IX.

*This book is known in the trade as "Antwerp I".

Update
Oct. 24th.

Modular Elliptic Curves and Fermat's Last Theorem

Andrew Wiles

(October 7, 1994)

Introduction

An elliptic curve over **Q** is said to be modular if it has a finite covering by a modular curve of the form $X_0(N)$. Any such elliptic curve has the property that its Hasse-Weil zeta function has an analytic continuation and satisfies a functional equation of the standard type. If an elliptic curve over **Q** with a given j-invariant is modular then it is easy to see that all elliptic curves with the same j-invariant are modular (in which case we say that the j-invariant is modular). A well known conjecture which grew out of the work of Shimura and Taniyama in the 1950's and 1960's ~~predicts~~ asserts that every elliptic curve over **Q** is modular. However, it only became widely known through its publication in a paper of Weil in 1967 in [We] (as an exercise for the interested reader!), in which moreover Weil gave conceptual evidence for it. Although it had been numerically verified in many cases, prior to the results described in this paper it had only been known that finitely many j-invariants were modular.

In 1985 Frey made the remarkable observation that this conjecture should imply Fermat's Last Theorem. The precise mechanism relating the two was formulated by Serre as the ϵ-conjecture and this was then proved by Ribet in the summer of 1986. Ribet's result only requires one to prove the conjecture for semistable elliptic curves in order to deduce Fermat's Last Theorem.

The first widely available manuscript

Notes on Fermat's Last Theorem

LECTURE XI

Recently, G. Frey has found a connection between Fermat's conjecture and elliptic curves, which so far seems to be the most compelling reason why the conjecture should be true. In short, if $a^p + b^p = c^p$ (p prime, a, b, c integers), consider the curve given by $y^2 = x(x - a^p)(x - c^p)$. Its discriminant is $4a^p b^p c^p$ and is much bigger than its conductor $2abc$. If we had a good effective Mordell theorem, we could derive that this cannot happen for big primes p. Also, if the curve were a Weil curve, we could find a modular form of weight 2 associated with its L-series. Ribet has shown that, modulo p, one can decrease the level of this form until it is one. However, there are no such forms. Thus Fermat's conjecture should follow if every elliptic curve over $\mathbb{Q}$ were a Weil curve, which is widely believed to be true, yet has not been demonstrated so far.

Gerd Faltings, 1991

It was long a mystery just why Euclid's geometry was as efficient as it seemed to be in describing our experience. Students busily learned how the real world could be extracted from a few axioms and the philosophers argued about whether the nature of reality as we knew it was *necessary*. Were the gods constrained to making our universe a mathematical one according to Euclid, as seemed to be? A sticking point was Euclid's fifth postulate, according to which exactly one line parallel to a given line passes through any point not on that line. Why not none, or several? The trouble was that it did not seem possible to show that anything other than Euclid's universe was feasible. Then the dam burst. Independently Gauss, Bolyai, and Lobachevsky showed that Euclid's principles, omitting that fifth axiom, of course, could readily be made to model geometries contradicting the fifth postulate. For example, one can take as points those of the upper half-plane $\mathcal{H}$, thus all $z = x + iy$ with $y > 0$, and as lines either vertical straight lines or semicircles with diameter parts of the real axis. By the way, the vertical straight lines are just such semicircles that happen to pass through infinity. Whatever, in this representation there are many "lines" through a point that are "parallel" to a given line, that is, that do not intersect the given "line".

In Euclidean geometry there are the obvious symmetries of translation, rotation, and reflection. In the *hyperbolic* geometry just described, translation is clearly once again a symmetry but so is inversion $S : z \mapsto -1/z$ in the unit circle. In fact, the maps S and $T : z \mapsto z+1$ together generate the complete group of linear fractional transformations $\Gamma = \mathrm{PSL}(2,\mathbb{Z})$ of 2×2 integer matrices with determinant 1, modulo $\{I, -I\}$. One notices that $S^2 = (TS)^3 = I$. These are the only relations. If one chooses to identify points that can be obtained one from the other by application of elements of Γ, then a *fundamental domain*, thus a domain consisting of exactly one representative of every point of $\mathcal{H}$, is given by the points $z = x + iy$, where, of course, $y > 0$ and $|z| > 1$ and with $-\frac{1}{2} \le x < \frac{1}{2}$, together with the boundary points on $|z| = 1$.

Now let me return to periodic functions. The idea is to remark that the map $x \mapsto e^{2i\pi x}$ defined *a priori* on the reals $\mathbb{R}$ is in practice defined on the space $\mathbb{R}/\mathbb{Z}$ — the reals modulo 1. Not only is it sufficient to know the function on the unit interval, but since the point 1 is equivalent to the point 0, one should think of the interval as joined at $0 \equiv 1$ thus forming a circle. One might therefore view the mapping as wrapping the real line onto the unit circle, with each single point on the circle corresponding to a complete *orbit* $\{x+n : n \in \mathbb{Z}\}$ of each point $x \in \mathbb{R}$. One talks about the circle being the *orbit space* and the mapping being the *covering map*.

In this spirit, let's now consider a doubly periodic mapping $z \mapsto \wp(z;\tau)$ defined a priori on the complex numbers $\mathbb{C}$. In this case the orbit space is the points of a parallelogram with vertices 0, 1, τ, and $1+\tau$ having opposite sides identified. Identifying one pair of sides thereby turns the parallelogram into the surface of a cylinder. Now identifying the other pair turns that surface into the surface of a *torus*, that is, of a doughnut. The role of $\mathbb{Z}$ in the first example is now played by the *lattice* of points $\Lambda = \{m\tau + n : m, n \in \mathbb{Z}\}$. One denotes the orbit space by $\mathbb{C}/\Lambda$, that is, the points of $\mathbb{C}$ modulo this lattice Λ. We already know that the Weierstrass $\wp$-function provides the covering mapping $\mathbb{C} \to \mathbb{C}/\Lambda$. Since this function is meromorphic on $\mathbb{C}$ it transfers the geometry of $\mathbb{C}$ across to the torus. Now remember that to be analytic is to have a Taylor expansion; meaning that locally, in arbitrarily small neighborhoods, the function is linear. Thus local geometry is preserved, which is to say that the angle between arcs is preserved under analytic maps. These mappings are said to be *conformal* — preserving form, at any rate angle. Such things as length of arc and area are not local notions; they are anything but preserved. With this induced geometry the torus becomes a *Riemann surface* having genus 1 as, among other things, it has just one hole. On the other hand, we know that any equation of the form $y^2 = 4x^3 - g_2x - g_3$ such that $g_2^3 - 27g_3^2 \neq 0$ — which is just saying that the cubic $4x^3 - g_2x - g_3$ has distinct zeros — models an elliptic curve. This curve is parametrized by elliptic functions $x = \wp(z;\tau)$ and $y = \wp'(z;\tau)$; in particular, it defines a lattice Λ in $\mathbb{C}$. In summary, there is a correspondence

$$\mathbb{C} \cup \{\infty\} \longrightarrow \text{elliptic curve}$$

wherein the points of the lattice correspond to the point O at infinity on the curve.

Just a quick by the way. We really should pause to ask just how and why the *genus* of a Riemann *surface* coincides with the *genus*, as explained in Lecture IX, of the *curve* defining the surface. A little bit of talk about differentials will do, much as George Szekeres taught me in my honors undergraduate year. But I'm going to leave it now, to enrich our sense of wonder and so that we have something yet to learn. The name of the relevant spell is the "Riemann-Roch Theorem".

An elliptic curve E has its Euclidean uniformization given by a mapping from $\mathbb{C}$, or rather $\mathbb{C} - \Lambda$, as just described. But there might also be a hyperbolic uniformization given by an appropriate mapping

$$\mathcal{H} - \{\text{a finite union of } \Gamma\text{-orbits}\} \longrightarrow E - \{\text{a finite set of points}\}.$$

It has been known for a while by a theorem of Bely that an elliptic curve admits a hyperbolic uniformization if and only if its invariants g_2 and g_3 are algebraic numbers. However we want more than just any hyperbolic uniformization.

Experience suggests that when it comes to arithmetic, the important subgroups are those defined by a congruence condition on the entries of the matrices comprising Γ. Unsurprisingly, such are known as *congruence subgroups*. Examples are the *principal congruence subgroups* $\Gamma_N = \Gamma(N)$ of level N consisting of those matrices

$$\begin{pmatrix} a & b \\ c & d \end{pmatrix} \equiv I \pmod{N}.$$

In other words, $a \equiv d \equiv 1$ and $c \equiv b \equiv 0 \pmod{N}$. Technically, the congruence subgroups of *level* N are those groups G such that

$$\Gamma(N) \subseteq G \subset \Gamma.$$

One easily obtains such groups just by selecting an integer matrix L of determinant N; then $\Gamma \bigcap L^{-1}\Gamma L$ provides a congruence group of level N. Taking $L = \left(\begin{smallmatrix} N & 0 \\ 0 & 1 \end{smallmatrix}\right)$, for example, provides the subgroup $\Gamma_0(N)$ defined in Γ by just the constraint $c \equiv 0 \pmod{N}$.

It's an interesting exercise to compute the *index* $\psi(N)$ of $\Gamma_0(N)$ in Γ, that is, the number of disjoint "translates" (technically *cosets*) of $\Gamma_0(N)$ required to cover Γ. One obtains

$$[\Gamma : \Gamma_0(N)] = \psi(N) = N \prod_{p|N} \left(1 + \frac{1}{p}\right).$$

Let me now define the space $Y_0(N) = \mathcal{H}/\Gamma_0(N)$, in effect the fundamental domain of $\Gamma_0(N)$, with geometry inherited from the upper half-plane $\mathcal{H}$. It always seems useful to "close", *compactify*, such spaces by including the *cusps* — in this case those points on the real line where the domain abuts the real axis, and the point $i\infty$ at infinity — to the space in order to obtain a Riemann surface, that is, to obtain the $X_0(N)$. Such compactification is quite analogous to our adding the point at infinity to the complex plane $\mathbb{C}$ to obtain the Riemann sphere. It turns $Y_0(N)$ into a Riemann surface, that is, into an algebraic curve. Oh! I'd better add a little more confusion by acknowledging that one will often find $\Gamma_0(N)\backslash\mathcal{H}$ written for $Y_0(N)$.

Barry Mazur* comments wisely that generations of students have been mystified by the fact that (algebraic) *curves* can be viewed as (Riemann) *surfaces*. He observes that the point is that if one is thinking algebraically, then $\mathbb{C}$ is just the affine line, thus a curve; whereas if one is thinking geometrically, then it is the complex plane, thus a surface.

From the present viewpoint, the content of the Taniyama-Weil-Shimura Conjecture — I may refer to it as just TWS from here on — is that a rational elliptic curve E always admits a hyperbolic uniformization with respect to some congruence subgroup of $\mathcal{H}$. A little more specifically: there is some positive integer N so that there is a nontrivial *morphism*, a structure-preserving map, defined over $\mathbb{Q}$ from the modular space $X_0(N)$ to E. Just as a correspondence $\mathbb{C}/\Lambda \longrightarrow E$ furnishes a parametrization of E using doubly periodic functions invariant under translations by the lattice Λ, so the modular representation provides a parametrization by automorphic (modular) functions invariant under action of elements of the group $\Gamma_0(N)$. Seen yet differently, we may view $\mathbb{C}/\Lambda$ as *being* (the set of complex points of) an elliptic curve. In that spirit we might expect to find E somehow as part of $X_0(N)$. Indeed one does, as a quotient of its Jacobian. But we don't need all that here. Suffice it to say as examples that if $X_0(N)$ happens to be of genus 1, then indeed, we may identify E and $X_0(N)$. And if $X_0(N)$ should have genus 0 we can say "Nonsense!". There cannot be a nontrivial morphism from such a space to an elliptic curve.

It was always clear that there is an algorithm allowing one, though with nontrivial difficulty, to compute all *modular* elliptic curves — Weil curves, as they were known until recently — arising from a given $X_0(N)$. Such Weil curves — one refers to them as having *analytic* conductor N — had a common property: a certain invariant which encodes the behavior of the curve when viewed as a cubic curve over the different finite fields†, their algebraic *conductor*, would turn out to be N.

Conversely, one could try to find all elliptic curves over $\mathbb{Q}$ of algebraic conductor N. That is a matter of solving certain diophantine equations. In principle such equations can always be solved. In practice doing so remains notoriously painful, except in lucky cases. For example, when

---

*The present remarks are in part my vulgarization of Barry Mazur's article 'Number theory as gadfly', *Amer. Math. Monthly* **98** (1991), pp. 593–610.

†See the Notes to Lecture IX.

there is no elliptic curve of the given conductor, then the corresponding diophantine equation has no integer solutions; and that it has no global solutions can often be seen by it having no solution modulo certain primes. So simple congruence arguments suffice. Thus the easiest cases in which to verify TWS were those for which there was known to be no modular elliptic curve of that conductor, typically certain cases with N prime, or of course with $X_0(N)$ of genus 0. It also turned out to be accessible to investigate all cases of conductor $N = 2^a 3^b$ and certain other cases where one could predict that the associated diophantine equation would be easy to solve by elementary methods. The first nontrivial case $N = 11$, in the context of which I learned the little I first knew, is harder.

In a more explicit formulation, suppose that $f(x, y) = 0$ defines a rational elliptic curve E. TWS claims directly that there are nonconstant modular functions g_x and g_y for the group $\Gamma_0(N)$ so that $f(g_x, g_y) = 0$. An argument that might require me to start talking about differentials on the Riemann surface $E(\mathbb{C})$ — the complex points on E — which I won't start here, shows that a certain combination derived from the functions g_x and g_y is a *cusp form* of *weight* 2 for the group $\Gamma_0(N)$. Recall that the phrase "cusp form" means that the Fourier expansion is such that this function is not just regular at cusps such as infinity, but actually vanishes there. More, using the fact that N is the conductor of E, one can show that there must exist a modular form as described, given by a Fourier expansion $a_1 q + a_2 q^2 + a_3 q^3 + a_4 q^4 + \cdots$, with all the a's *integers*, and $a_1 = 1$. The amazing thing is that it turns out that these integers a_n, which arise from analytic and algebraic geometric properties of $X_0(N)$, have to be — at any rate for $n = p$ prime — exactly the $a_p = p + 1 - N_p$ that arise from counting the number of points N_p on the curve E viewed as defined over the finite field F_p. Recognizing this arithmetic interpretation is essential to the eventual strategy for proving TWS.

Of course, it's absurd to keep on talking about the conjecture without remarking on the names used for it. In 1955 Yutaka Taniyama asked a somewhat more vague and rather more general question about relationships between elliptic curves defined over arbitrary number fields and certain automorphic forms. Subsequently, André Weil's work of 1967 on Dirichlet series, generalizing that of Hecke to the congruence subgroups, created the context in which I first met the conjecture. The question was: Is every rational elliptic curve a Weil curve? Weil acknowledged that this supposition "problematisch scheint". The historically well informed then talked about the Taniyama-Weil Conjecture. In the meantime, following on work of Eichler, Goro Shimura's work had led to an arithmetic description of the Fourier coefficients a_n above, a detailed understanding of requirements to establish the conjecture, and its proof for CM-curves — elliptic curves with complex multiplication.

Notes and Remarks

XI.1 The quotation with which I open this lecture is from 'Recent progress in diophantine geometry', by Gerd Faltings, in *Mathematical Research Today and Tomorrow*, C. Casacuberta and M. Castellet eds. (Symposium on the Current State and Prospects of Mathematics, Barcelona, June 1991) Lecture Notes in Mathematics **1525** (New York: Springer-Verlag, 1992).

XI.2 Peter Merrotsy, one of the readers of my draft, asked me to explain why Faltings can claim that "the curve given by $y^2 = x(x - a^p)(x - c^p)$ [has] discriminant $4a^p b^p c^p$ [which] is much bigger than its conductor $2abc$." This is the closing observation in a survey and has a little of the looseness of some of my remarks. On the surface it's quite correct. The model does have discriminant $4a^p b^p c^p$ according to the Notes to Lecture IX; then the apparent conductor is indeed $2abc$. Of course, the stated model may not be minimal. However, one might remark that for large p such a discriminant and conductor entail that Mordell's equation $x^3 - y^2 = k$ has a solution in integers x and y for a k very much larger than its "conductor" (see Lecture XIV) — the product of the distinct primes dividing it. That runs against the prevailing philosophy on the existence of solutions of such diophantine equations. Mind you, Faltings goes on to remark that "if we had a good effective Mordell Theorem, we could derive that this cannot happen for big primes p". Here he is asking *inter alia* (see Lecture XV) that we be able to actually find the rank of any given elliptic curve, which somehow seems a yet bigger ask.

XI.3 A detailed but easy-to-read introduction to the development and history of noneuclidean geometry can be found in Marvin Jay Greenberg, *Euclidean and Non-Euclidean Geometries* (San Francisco: W. H. Freeman, 1972).

XI.4 I first learned about Riemann surfaces from George Springer's book, appropriately called *Introduction to Riemann Surfaces* (Reading, Mass.: Addison-Wesley, 1957); now reprinted (New York: Chelsea, 1981). One could also look at Hermann Weyl's *The Concept of a Riemann Surface*, (Reading, Mass.: Addison-Wesley, 1955). A principal object of study for such things is the Riemann-Roch theorem, which ties together much of the geometry and algebra of the situation. It is such unifying results that are at the very core of mathematics. The Taniyama-Wiles theorem is of this kind.

XI.5 A primary trick in calculating $\psi(N)$ is to observe that in fact ψ is a *multiplicative* function: if $\gcd(M, N) = 1$, then $\psi(M)\psi(N) = \psi(MN)$. Thus it suffices to know just the $\psi(p^r)$. This is not at all a profound observation, since it is just the Chinese Remainder Theorem applied to the equation $ad - bc \equiv 1 \pmod{N}$, subject to $N \mid c$. It's useful to begin by calculating the index of the principal congruence subgroup $\Gamma(N)$ in Γ. So suppose that $ad - bc \equiv 1 \pmod{p^\alpha}$. If $a \not\equiv 0 \pmod{p}$, there are $\phi(p^\alpha)$ possibilities for a; we may choose b and c arbitrarily making $p^{2\alpha}$ choices, and now d is determined. Otherwise, for the remaining $p^{\alpha-1}$ choices for a, d is arbitrary and now $\phi(p^\alpha)$ choices for, say, b determines c. All together, there are

$$\phi(p^\alpha)p^{2\alpha} + p^{2\alpha-1}\phi(p^\alpha) = p^{3\alpha}\left(1 - \frac{1}{p^2}\right)$$

possibilities. It follows that the index of $\Gamma(N)$ in Γ is

$$N^3 \prod_{p|N}\left(1 - \frac{1}{p^2}\right).$$

If, on the other hand, $N|c$, then b is arbitrary and $a \not\equiv 0 \mod N$ thus determines d. We must divide the previous result by $N\phi(N)$, obtaining

$$\psi(N) = N\prod_{p|N}\left(1 + \frac{1}{p}\right),$$

as asserted.

XI.6 I really should be a bit more careful, so let me say, or say again, what is meant by a modular function for the group $\Gamma_0(N)$. It is a function $f : \mathcal{H} \to \mathbb{C}$, acting on the upper half-plane $\mathcal{H}$ and taking values in $\mathbb{C}$, which is invariant under the action of $\Gamma_0(N)$ on $\mathcal{H}$. Thus

$$f\left(\frac{az+b}{cz+b}\right) = f(z) \qquad \text{for all integers } a, b, c, d : ad - bc = 1 \text{ and } N|c.$$

Now comes the bit I've been neglecting. It is part of the definition of "modular function" that f be *meromorphic at the cusps.* For example, it is easy to see that f is periodic with period N. The demand that it be meromorphic at the cusp ∞ at infinity then is that its Fourier expansion

$$\sum_{-\infty}^{\infty} c_n q^{n/N}$$

in fact have only *finitely* many terms with negative n; that is, $c_{-m} = 0$ for all m greater than some M. Similarly, the cusp at a rational a/b becomes a cusp at infinity of $g(z) = f((az+b)/(cz+d))$, which must therefore also have a Fourier expansion with the property just described.

XI.7 Let's see just how modular forms of weight 2 appear. Suppose that the elliptic curve $y^2 = x^3 + Ax + B$ is parametrized by modular functions; meaning that we have

$$(g_y(z))^2 = (g_x(z))^3 + Ag_x(z) + B.$$

The elliptic integral of the first kind

$$\int \frac{dx}{\sqrt{x^3 + Ax + B}} = \int \frac{dx}{y}$$

might move one to write

$$\frac{dx}{y} = \frac{dg_x}{g_y} = \frac{g'_x(z)\,dz}{g_y(z)} = G(z)\,dz\,.$$

Because g_x and g_y are invariant under transformations from the group $\Gamma_0(N)$, it's now a simple calculation to confirm that

$$G\left(\frac{az+b}{cz+d}\right) = (cz+d)^{-2}G(z) \quad \text{for} \quad \begin{pmatrix} a & b \\ c & d \end{pmatrix} \in \Gamma_0(N)\,.$$

XI.8 An example due to Ligozat* states that the curve E given by

$$y^2 + xy + y = x^3 + x^2$$

is parametrized by $X_0(15)$. The corresponding cusp form is

$$q\prod_{n=1}^{\infty}(1-q^n)(1-q^{3n})(1-q^{5n})(1-q^{15n})\,.$$

Since the coefficient of q^7 is 1, it follows that there are precisely $7 - 1 = 6$ solutions to the equation modulo 7, that is six finite points on the curve E over $\mathbb{F}_7$.

Oops!? you say. Shouldn't that be $(7+1) - 1 = 7$? But no. Sure the reduction at 7 of the curve has seven rational points, but only six of those are finite. So the equation modulo 7 has just six *solutions*.

XI.9 The elliptic curve given by by $y^2 = x^3 - x$ is parametrized by $X_0(32)$. The corresponding cusp form is

$$q\prod_{n=1}^{\infty}(1-q^{4n})^2(1-q^{8n})^2\,.$$

In the power series expansion only terms with exponent $\equiv 1 \pmod 4$ occur. Indeed, if $p \equiv 3 \pmod 4$, then the number of points over $\mathbb{F}_p$ on the curve is exactly $p+1$. One for the point at infinity, one for each of $x = 0, \pm 1$, and in addition if $a^3 - a$ is not a square, then $(-a)^3 - (-a)$ is; moreover, -1 is not a square in $\mathbb{F}_p$ and the product of two nonsquares is a square. In total that yields $4 + 2(p-3)/2 = p+1$ points. Thus indeed $a_p = p + 1 - (p+1) = 0$ for these primes.

Moreover, if $p \equiv 1 \pmod 4$, then p has a unique representation as $p = a^2 + 4b^2$. Now choose the sign of a so that $a \equiv 1 \pmod 4$ if b is even and $a \equiv -1 \pmod 4$ if b is odd. It is then easy to see that $a_p = 2a$ and to confirm directly, say for the examples $p = 5, 13, \ldots$, that the number of solutions of the equation $y^2 = x^3 - x$ modulo p is indeed $p - 2a$.

XI.10 I am indebted for these examples and remarks to a survey by Jaap Top, 'Hoe bewijst Wiles het Taniyama-Weil vermoeden?', emanating from the Dutch "Fermatdag", already alluded to in the Notes to Lecture VI. Jaap

---

*Gérard Ligozat, 'Courbes modulaires de genre 1', Mém. Soc. Math. France **43** (1975).

further remarks on the role played by the Isogeny Theorem of Serre and Faltings, whereby two elliptic curves E and E' have the same L-function if and only if there's a surjective homomorphism $E \to E'$ defined over $\mathbb{Q}$. The point is that there is an injective map from $X_0(N)$ into its Jacobian variety, and a surjective map from that Jacobian to its every abelian factor. With the help of Hecke operators one finds such a map to a factor E, and then general considerations yield a surjective morphism $X_0(N) \to E$.

XI.11 No sooner had I written much of the final paragraph of the Lecture that I came to see a letter written by Serge Lang suggesting forcefully that the Modularity Conjecture must be called Taniyama-Shimura by all decently informed people. I should add that Ribet, in his Prix Fermat lecture 'From the Taniyama-Shimura conjecture to Fermat's Last Theorem' [*Ann. Fac. Sc. Toulouse*], refers just to Taniyama and Shimura. Moreover, Shimura (for example, in a letter to Ribet) recalls having been quite explicit c.1962–1966 in suggesting in conversation that every elliptic curve over $\mathbb{Q}$ is modular, allegedly startling Serre, Weil, and others with the claim. At that time Taniyama's questions were quite forgotten.*

The trouble is that in this, as in all other matters, even the testimony of the participants is not all that reliable. In any case, Shimura's influence on the subject of the conjecture is well recognized. It certainly seems strange to rewrite history so as to deny a rôle to Weil, whose paper was seminal at least in convincing the world of the likelihood of the truth of the matter. Indeed, throughout the seventies the conjecture seemed entirely blamed on Weil. Serre's suggestion, made at the Hong Kong meeting on "Elliptic functions and modular forms" (December 1993), that we refer mysteriously to "The $***$-Conjecture", has considerable wisdom. Much better yet, one refers to the "Modularity Conjecture". Then the name describes the actual content of the conjecture and avoids silly jokes, all too often true of the named theorems in the undergraduate curriculum, whereby one was told that if a theorem had two names attached to it, then invariably the first name does not belong at all.†

XI.12 In any case, there is a tradition in mathematics of misnaming things. A celebrated example is "Pell's equation" $X^2 - DY^2 = 1$ — I should be showing decent distaste by speaking of the "so-called Pell's equation". A proper Eurocentric viewpoint (the problem was known to Indian mathematics way back) would have us confusingly say "Fermat's equation". For a while textbooks officiously, but absurdly, called it the "Pell-Fermat equation" (and no doubt, "Fermat-Pell equation", as well).

---

*I recall the forgetting. When working with Coates on verifying the Weil conjecture for elliptic curves of conductor 11, we called it the Weil Conjecture at all stages until, right towards the end, Coates advised me of the news that one now spoke of the "Taniyama-Weil" Conjecture.

†Extended to joint papers this is just a calumny. There is no particular reason for the person whose name comes first in the alphabet to have been the author that barely contributed to the paper at all.

Several years ago I had the happy experience of being rung by the secretary of a colleague at Sydney University to be told that "Prof" was writing a history of mathematics at the university. Apparently a Professor Pell had been head of the department late last century. "Prof" was wondering whether that Pell was the Pell of Pell's equation. I was able to say with confidence that this was not so, both on the grounds that the matter went back to the seventeenth century and because it was well known that Euler had made a mistaken attribution in his textbook; no Pell had had anything to do with Pell's equation. Nevertheless, I have referred, and will continue to refer, to Pell's equation with complete equanimity as "Pell's equation".

And *pace* Serge Lang, one might recall such things as Tate explaining ['The arithmetic of elliptic curves', *Invent. Math.* **23** (1974), pp. 179–206] that one of his topics is "Modular curves and Weil's astounding idea that every elliptic curve over the rational field is 'modular'." So it will take a while before it will seem right to cease mentioning Weil in the context of the $***$-Conjecture.

XI.13 Elsewhere I use the name "Modularity Conjecture". I was pleased at my word play with "a while" and "Weil" above (though, as an e-mail discussion on the pronunciation reminded me, it is pronounced more like the "vei(l)" in "oy veh", by the way). And I was delighted when at an Oberwolfach meeting in May 1995, Larry Washington referred to the "Taniyama–Wiles Theorem".

XI.14 Modularity, in the sense that an elliptic curve is associated to a modular cusp form of weight 2 as briefly hinted at in my remarks, is essentially confined to elliptic curves defined over $\mathbb{Q}$. If an elliptic curve E, defined over some extension field of $\mathbb{Q}$ hopes to be modular, then it is at any rate necessary that its conjugate curves E^σ be isogenous to E. It is a generalization of TWS that also such "rational sets" of isogenous elliptic curves are always modular.

XI.15 Michel Mendès France reminds me to tell the story of Bombieri's napkin. At the Queen's University number theory meeting in 1979, Roger Apéry was a victim of Enrico Bombieri's observation that "the equation

$$\binom{x}{n} + \binom{y}{n} = \binom{z}{n}$$

has no trivial solutions for $n \geq 3$." At breakfast, next morning, Apéry excitedly reported having spent the night finding the smallest example $x = 10$, $y = 16$, $z = 17$, with $n = 3$. "Just so", responded Bombieri. "I said there was no trivial solution!"

LECTURE XII

Since we're mathematicians, we all know that $\int_{-\infty}^{\infty} e^{-\pi t^2} dt = 1$, for this is the content of Lord Kelvin's dictum that "a mathematician is a person to whom

$$\int_{-\infty}^{\infty} e^{-x^2} dx = \sqrt{\pi}$$

is as obvious as $1 + 1 = 2$."* Let's see a consequence of this happy fact.

Given a suitable function f, we've noticed that

$$g(x) = \sum_{-\infty}^{\infty} f(x+n)$$

is blatantly periodic with period 1. But if g is periodic we should make it admit to that by displaying its Fourier expansion

$$g(x) = \sum_{-\infty}^{\infty} \Big(\int_0^1 g(u) e^{-2i\pi ku} \, du \Big) e^{2i\pi kx} .$$

The fun thing is that assuming that we may interchange integration and summation,

$$\int_0^1 g(u) e^{-2i\pi ku} \, du = \int_0^1 \sum_{-\infty}^{\infty} f(u+n) e^{-2i\pi ku} \, du = \int_{-\infty}^{\infty} f(u) e^{-2i\pi ku} \, du .$$

So

$$\sum_{-\infty}^{\infty} f(n) = \sum_{k=-\infty}^{\infty} \int_{-\infty}^{\infty} f(u) e^{-2i\pi ku} \, du = \sum_{k=-\infty}^{\infty} \hat{f}(k) .$$

The last equality is just the definition of the *Fourier transform* $\hat{f}$ of f, made so that I display the *Poisson summation formula* in its full glory.

*I once found myself actually tested on this point. In my then capacity of Chairman of Directors of University Co-operative Bookshop Limited — which operates a chain of college and university bookshops on the Australian east coast — I presided at the launching of a book by Professor Julius Sumner Miller, a populist scientist. I introduced him and admitted to being a mathematician in real life. Waxing eloquent in his speech featuring Lord Kelvin, he suddenly turned upon me with the question, waited for me to hesitantly mutter "$\sqrt{\pi}$", and turned back to his audience with "You see, he *is* a mathematician!"

Just about the simplest admissible case is $f(x) = e^{-\pi t x^2}$. Then by Lord Kelvin,

$$\hat{f}(k) = \int_{-\infty}^{\infty} e^{-\pi t u^2 - 2i\pi k u} du = \sqrt{\frac{1}{t}}\, e^{-\pi k^2/t}.$$

So *mirabile dictu*,

$$\sum_{-\infty}^{\infty} e^{-\pi t n^2} = \sqrt{\frac{1}{t}} \sum_{-\infty}^{\infty} e^{-\pi n^2/t}.$$

The good thing about this example is that it corresponds to the simplest zeta function, the Riemann ζ-function $\zeta(s) = \sum_{n=1}^{\infty} n^{-s}$. The correspondence is by way of the Mellin transform, which I briefly mentioned in Lecture X. The idea is this. One defines the Γ-function by

$$\Gamma(s) = \int_0^{\infty} e^{-x} x^{s-1} dx.$$

Then

$$\int_0^{\infty} x^{s/2-1}\left(\sum_{n=1}^{\infty} e^{-\pi n^2 x}\right) dx = \zeta(s)\Gamma(\tfrac{1}{2}s)\pi^{-s/2}.$$

It now turns out that the functional equation for the ϑ-*function*, namely for $\vartheta(z) = \sum_{-\infty}^{\infty} e^{\pi i n^2 z}$ just shown to us by Poisson summation, leads to a functional equation for the Riemann ζ-function. This is not terribly subtle. We simply split the interval of integration into the two parts $[0, 1]$ and $[1, \infty)$ and apply the transformation for the first interval. That yields

$$\zeta(s)\Gamma(\tfrac{1}{2}s)\pi^{-s/2} = -\left(\frac{1}{1-s} + \frac{1}{s}\right) + \int_1^{\infty} (t^{s/2} + t^{(1-s)/2})\left(\sum_{n=1}^{\infty} e^{-\pi n^2 t}\right) dt,$$

and we see, by symmetry, that the function $\zeta(s)\Gamma(\frac{1}{2}s)\pi^{-s/2}$ is invariant under the transformation $s \mapsto (1 - s)$.

We're applying a fundamental principle of function theory when we say that the functions $(1 - z)^{-1}$ defined for $z \neq 1$, and $\sum_{n=0}^{\infty} z^n$ defined for just $|z| < 1$, represent the *same* function. If two analytic functions coincide for some region then they represent one and the same function. The trouble is that the functions $\Lambda(s) = \zeta(s)\Gamma(\frac{1}{2}s)\pi^{-s/2}$ and $\Lambda(1 - s)$ do not obviously have a common region. As usual I write $s = \sigma + it$. Naïvely, $\zeta(s)$ is just defined for $\sigma > 1$; and for that matter, the integral defining $\Gamma(\frac{1}{2}s)$ converges just for $\sigma > 2$. However, that's not significant because integration by parts readily yields $s\Gamma(s) = \Gamma(s + 1)$, providing an analytic continuation into the left half-plane for the Γ-function. Indeed, well-brought-up people know that

$$\frac{1}{\Gamma(s)} = s\, e^{\gamma s} \prod_{n=1}^{\infty} \left(1 + \frac{s}{n}\right) e^{-s/n},$$

where $\gamma = \lim_{n\to\infty}(1 + \frac{1}{2} + \frac{1}{3} + \cdots + \frac{1}{n} - \log n)$ is the Euler constant. This shows that the Γ-function $\Gamma(s)$ is a meromorphic function that never vanishes and has simple poles at $s = 0, -1, -2, \ldots$. If we recall the product for $\sin \pi z$ of Lecture VIII we also see that

$$\Gamma(s) \cdot (-s\Gamma(-s)) = \Gamma(s)\Gamma(1-s) = \pi s / \sin(\pi s).$$

As regards the ζ-function, the trick turns out to be to notice that

$$\zeta(s) - \frac{1}{s-1}$$

is in fact holomorphic for $\sigma > 0$. Indeed, since

$$\frac{1}{s-1} = \int_1^\infty t^{-s}\,dt = \sum_{n=1}^\infty \int_n^{n+1} t^{-s}\,dt,$$

we have

$$\zeta(s) = \frac{1}{s-1} + \sum_{n=1}^\infty \int_n^{n+1} (n^{-s} - t^{-s})\,dt.$$

That provides an analytic continuation to $\sigma > 0$ and now the functional equation extends the Riemann ζ-function to a meromorphic function on the entire plane, with just the simple pole at $s = 1$ as its only singularity. The poles of the Γ-function show that $\zeta(s)$ has zeros, its so-called *trivial zeros*, at $s = 0, -2, -4, \ldots$.

The point of this old story is that starting with the theta function, given as $\vartheta(z) = \sum_{-\infty}^{\infty} e^{\pi i n^2 z}$ defined on the upper half-plane and satisfying

$$\vartheta(z+2) = \vartheta(z) \quad \text{and} \quad \vartheta(-1/z) = \sqrt{z/i}\,\vartheta(z),$$

so that it is a modular form of weight $\frac{1}{2}$ for the group generated by $z \mapsto z+2$ and $z \mapsto -1/z$, we obtain a functional equation

$$\zeta(s)\,\Gamma(\tfrac{1}{2}s)\,\pi^{-\frac{1}{2}s} = \zeta(1-s)\,\Gamma(\tfrac{1}{2}(1-s))\,\pi^{\frac{1}{2}(s-1)}$$

for the corresponding Dirichlet series $\zeta(s)$. This argument can be run in reverse. There is an inverse Mellin transform allowing one to move from a functional equation for a Dirichlet series to obtain a corresponding modular function.

Hecke shows, among a great many related things, that the phenomenon just described is much more general. That is a good thing because it allows one to move from Dirichlet series with functional equation to modular forms. Modular forms are rather easier to handle than Dirichlet series. For one thing, the modular forms for the group Γ of the same weight k obviously form a vector space. That seems trite to notice, but of course Hecke works this out in detail. The endomorphisms of a vector space — the linear maps of the space into itself — provide an algebra of operators. This Hecke algebra gives information about the vector space and thence about the original Dirichlet series.

In fact, this is what this Taniyama-Weil-Shimura business is all about. The L-function associated with a rational elliptic curve is a Dirichlet series detailing the arithmetic of E. Taniyama raised the suspicion that now and then this function might correspond to a modular form. By work of Eichler and of Shimura one began to see how such a modular form might arise in the first place. It would be a cusp form of weight two for the group $\Gamma_0(N)$, some integer N. Hecke's arguments were carried over to the case of level N by Weil; now all was clear. The integer N, the so-called analytic conductor of E, should coincide with the algebraic conductor of E — that latter conductor being a number describing the bad reductions of the curve E. Following Weil, one could start with the modular curve $X_0(N)$, the compactification of $\mathcal{H}/\Gamma_0(N)$ — the upper half-plane modulo the action of the group $\Gamma_0(N)$, and find the elliptic curves arising from $X_0(N)$. At this point the question becomes whether all rational elliptic curves are Weil curves. Certainly, Weil curves are *good*. Because one has the cusp form to start with, one knows that the corresponding L-function has a functional equation; and one knows that equation. In particular, it now makes sense to study the value $L(E,1)$ and to conjecture that it reveals the nature of the group $E(\mathbb{Q})$ of rational points on E.

The case that is most accessible is that of elliptic curves with Complex Multiplication.* This notion derives from an elementary property of the period lattice $\tau\mathbb{Z} + \mathbb{Z}$ of the elliptic function parametrizing the curve. Obviously, a nonzero integer n defines a sublattice $n\tau\mathbb{Z} + n\mathbb{Z}$ which is n^2 times as coarse as the original lattice. Plainly, n cannot be replaced by any fractional number if we are to obtain a sublattice. However, consider an arbitrary α not a rational integer, so that for integers a, b, c, and d

$$\alpha \cdot \tau = a\tau + b \quad \text{and} \quad \alpha \cdot 1 = c\tau + d\,.$$

Then α is an eigenvalue of the matrix $\begin{pmatrix} a & b \\ c & d \end{pmatrix}$, and if it's not in $\mathbb{Z}$, then it is a quadratic irrational algebraic integer. Since τ is dinkum complex and $\alpha = c\tau + d$, it follows that α and τ must be *complex* quadratic irrational integers, hence CM.

The extra structure that complex multiplication provides makes all sorts of things accessible. Indeed, in this case the analogy with the known number field case was recognized to be quite close; the functional equation for the L-function of elliptic curves was already known to Deuring. Shimura could show that rational elliptic curves with CM are modular — thus his $*$ in the $***$-Conjecture. And in 1977 Coates and Wiles† proved the weakest instance of the Birch–Swinnerton-Dyer Conjecture, namely that the L-function of a rational elliptic curve with CM vanishes at $s = 1$ if it has infinitely many rational points.

---

*The capitalization is to allude to the universal abbreviation "CM".

†Yes, the same Andrew Wiles, of course. Wiles did not spring fresh out of the shell with his proof of the FLT. I was scandalized some years ago when my then Ph.D. student, Deryn Griffiths, mentioned that she had a cousin in England who was some sort of mathematician. His name was Wiles. Had I heard of him?

Every quadratic irrational integer τ gives rise to an elliptic curve with CM. But we want just rational elliptic curves, thus $y^2 = 4x^3 - g_2x - g_3$ with *rational* g_2 and g_3, so with $j(\tau)$ rational.

The beautiful fact is that if τ is a quadratic irrational integer then $j(\tau)$ is an algebraic integer. And this integer is rational, that is in $\mathbb{Z}$, if and only if the *order* $\tau\mathbb{Z} + \mathbb{Z}$ has unique factorization. Such matters have been studied for some two hundred years. Gauss made calculations suggesting that the only complex quadratic fields $\mathbb{Q}(\sqrt{-D})$ with class number one are those with $D = 1, 2, 3, 7, 11, 19, 43, 67$, and 163. By the thirties, Heilbronn could show that there is at most one other case, and that there are no more was proved by Heegner in 1952. Heegner's paper was poorly presented and was not recognized as essentially correct until the class number one problem had been settled by different means independently by Alan Baker and by Harold Stark in the mid-sixties. Because these fields provide four additional cases, by way of nonmaximal orders of class number one, there are 13 values $j(\tau)$ in all corresponding to rational elliptic curves with CM.

Notes and Remarks

XII.1 The simplest way of becoming a mathematician is to remember polar coordinates and to notice that

$$\left(\int_{-\infty}^{\infty} e^{-t^2}dt\right)^2 = \left(\int_{-\infty}^{\infty} e^{-x^2}dx\right)\left(\int_{-\infty}^{\infty} e^{-y^2}dy\right) =$$

$$= \int_{-\infty}^{\infty}\int_{-\infty}^{\infty} e^{-(x^2+y^2)}dxdy = \int_0^{2\pi}\int_0^{\infty} e^{-r^2}r\,drd\theta = \pi\,.$$

XII.2 Nowadays, anyone learning the $\mathcal{A}\mathcal{M}\mathcal{S}$-TEX variant of the TEX typesetting language can readily become a mathematician. Lord Kelvin's remark is the content of Michael Spivak's first example in his *The Joy of TEX*, 2nd edition (Providence, R.I.: American Mathematical Society, 1993).

XII.3 It's a matter of taste and emphasis really but here is another, rather different argument telling Riemann's story of the θ-"constant" ϑ and its modularity. Set

$$\theta_3(z \mid \tau) = \sum_{n\in\mathbb{Z}} e^{2i\pi\left(nz+\frac{1}{2}n^2\tau\right)}$$

and notice that $\theta_3(z \mid \tau)$ is an entire function of z, with plainly

$$\theta_3(z+1 \mid \tau) = \theta_3(z \mid \tau) \quad \text{and} \quad \theta_3(z+\tau \mid \tau) = e^{-i\pi(2z+\tau)}\,\theta_3(z \mid \tau)\,.$$

Moreover the zeros of $\theta_3(z \mid \tau)$ are simple. They are located at the values $(m+\frac{1}{2})+(n+\frac{1}{2})\tau$ for $m, n \in \mathbb{Z}$. Viewing θ_3 as a function of both variables, we have the *heat equation*

$$\frac{\partial^2 \theta_3}{\partial z^2} = 4i\pi \frac{\partial \theta_3}{\partial \tau}.$$

Now the function $\psi(z \mid \tau) = \theta_3\left(\frac{z}{\tau} \,\middle|\, \frac{-1}{\tau}\right)$ has the same zeros as θ_3 and by "periodicity" $\psi(z \mid \tau) = g(\tau)e^{i\pi z^2/\tau}\theta_3(z \mid \tau)$. The heat equation shows that $g(\tau) = k\sqrt{\tau/i}$. Checking at $\tau = i$ yields the value of the constant as $k = 1$. Hence

$$\theta_3\left(\frac{z}{\tau} \,\middle|\, \frac{-1}{\tau}\right) = \sqrt{\tau/i}\, e^{i\pi z^2/\tau}\, \theta_3(z \mid \tau).$$

Now simply put $z = 0$. This is the more traditional version, and I might well have extracted the argument from E. T. Whittaker and G. N. Watson, *A Course of Modern Analysis* (Cambridge: Cambridge University Press, 1927). Notwithstanding its title, *Whittaker and Watson* is neither "modern" nor "analysis", as we now understand it. But it is the bible of the classical special functions. In the event, however, I picked the present details from Robert Gergondey, 'Decorated elliptic curves: modular aspects', in M. Waldschmidt et al. eds., *From Number Theory to Physics* (New York: Springer–Verlag, 1992), Chapter 5.

XII.4 I remark mainly as a bit of advertising that other than for its value, not much is known about the Euler constant γ. For all we can prove, the Euler constant might be a rational number (though of course we "know" that it surely is transcendental). Until relatively recently the situation with $\zeta(3)$ was the same. Then Apéry gave a quite astonishing proof of its irrationality. For the story, see my 'A proof that Euler missed ... Apéry's proof of the irrationality of $\zeta(3)$; An informal report'[*Math. Intelligencer* **1** (1979), pp. 195–203]. We still cannot prove anything about the status of $\zeta(5), \zeta(7), \ldots$. Of course, we know that $\zeta(2), \zeta(4), \ldots$ are transcendental by virtue of the well-known transcendence of π.

I feel moved to sing the praises of my report, but let me leave it to Paul Halmos to suggest that it's not too bad. I was hugely cheered back then to read a letter from Halmos [*Notices Amer. Math. Soc.* **30** (1983), pp. 600–601] berating the American Mathematical Society for the nature of the research-expository articles in the *Bulletin of the American Mathematical Society*. Halmos goes on to say: "Can bona fide research-expository articles really exist? Have there been any, anywhere, any time? I feel the answer must be yes, but I cannot think of an example that everyone would find convincing. Is van der Poorten's report on the irrationality of $\zeta(3)$ a possibility?"

Had Halmos waited one more year he could have cited Don Zagier's article 'L-series of elliptic curves, the Birch–Swinnerton-Dyer Conjecture, and the class number problem of Gauss', [*Notices Amer. Math. Soc.* 31 (1984), pp. 739–743] as a certainty. It forms the basis of Lecture XVI below.

XII.5 Since the derivative of the function $n^{-s} - t^{-s}$ is s/t^{s+1}, it follows that

$$\sup_{n \le t \le n+1} |n^{-s} - t^{-s}| \le |s|/n^{\sigma+1},$$

making it plain, given the remarks made above, that indeed in $\sigma > 0$ the only singularity of $\zeta(s)$ is the simple pole at $s = 1$.

XII.6 In the case of the L-series $L(E, s)$ belonging to an elliptic curve E defined over $\mathbb{Q}$, it is a consequence of the Modularity Conjecture that there is a positive integer N, and an $\varepsilon = \pm 1$, so that having set

$$\Lambda(E, s) = N^{\frac{1}{2}s}(2\pi)^{-s}\Gamma(s)L(E, s)$$

we have the functional equation

$$\Lambda(E, 2 - s) = \varepsilon\Lambda(E, s).$$

Notice that when $\varepsilon = -1$, then $L(E, 1) = 0$, if it exists. Incidentally, one expects that if r is the rank of the group $E(\mathbb{Q})$ of rational points (recall that the *rank* is the number of independent generators for the points of infinite order), then $\varepsilon = (-1)^r$. These and related considerations underlie the prediction of Birch and Swinnerton-Dyer, that $L(E, 1) = 0$ if and only if $E(\mathbb{Q})$ is infinite. Mind you, in general, one only knows *a priori* that the L-series is defined for real part $\Re(s) > \frac{3}{2}$, so any chat about the L-function at 1 at the least requires assumptions entailing the possibility to analytically continue the L-series to the left of $s = \frac{3}{2}$.

For elliptic curves E with complex multiplication these matters are a theorem of Deuring; and in that CM case Shimura showed the truth of the Modularity Conjecture. Still assuming CM, Coates and Wiles confirmed that indeed if $E(\mathbb{Q})$ is infinite, then $L(E, 1) = 0$.*

XII.7 The integer N is called the *analytic conductor* of the curve E. As the notation and terminology suggest, it is part of the web of conjectures concerning elliptic curves over $\mathbb{Q}$ that this analytic conductor N coincides with the algebraic conductor N mentioned in the Notes to Lecture IX.

XII.8 It doesn't seem profound to remark that given some group Γ of transformations, the modular forms of given weight k form a vector space $M(\Gamma, k)$, and that the cusp forms of that weight form a subspace $S(\Gamma, k)$ of $M(\Gamma, k)$. After all, this says no more than that linear combinations of such forms are again such forms. Still, we might be led to look for extra structure on these spaces. Are there linear transformations of these spaces into themselves that are somehow sympathetic to the underlying arithmetic? And is there an inner product, so that those endomorphisms also respect the metric structure then provided?

---

*I should confess that only the intervention of Jaap Top prevented me from claiming an erroneous "iff" here and earlier. The sad fact is that not even Kolyvagin can as yet show that $L(E, 1) = 0$ entails that there are infinitely many rational points on E. The best we know follows from Gross-Zagier, whereby $L(E, 1) = 0$ *and* $L'(E, 1) \neq 0$ implies $E(Q)$ infinite, at any rate for modular elliptic curves. See Lecture XVI.

It turns out that one should begin by studying certain correspondences on lattices. Indeed, take the correspondences $T(n)$ which associate to a lattice Ω its sublattices Ω' of index n in Ω. We might write

$$T(n)(\Omega) = n^{k-1} \sum_{\substack{\Omega' \subseteq \Omega \\ [\Omega:\Omega']=n}} \Omega'$$

as a formal way of indicating that each of the sublattices Ω' occurs once. The multiplier n^{k-1} turns out to be convenient when eventually we induce a linear map on modular forms of weight k. One can then describe the consecutive action of the Hecke operators $T(m)$ and $T(n)$ by

$$T(n)T(m) = \sum_{d|n,m} d^{k-1} T\left(\frac{nm}{d^2}\right).$$

The interrelationships between the actions is summed up by the identity

$$\sum_{n=1}^{\infty} T(n)n^{-s} = \prod_{p} (I - p^{-s}T(p) + p^{k-1-2s}I)^{-1},$$

where I is the identity action.

The usual way to describe the induced operations on modular forms F of weight k is as follows. For $\gamma = \left(\begin{smallmatrix} a & b \\ c & d \end{smallmatrix}\right)$ write

$$F_{|\gamma}(z) = (ad - bc)^{k/2}(cz + d)^{-k} F\left(\frac{az + b}{cz + d}\right).$$

Then $T(n)$ induces a linear transformation

$$T_n F = n^{\frac{1}{2}k-1} \sum_{\mu \in \mathcal{M}_n/\Gamma} F_{|\mu}$$

with the sum over representatives μ of 2×2 integer matrices of determinant n, modulo the action of Γ. In concrete terms this is

$$T_n F(z) = n^{k-1} \sum_{\substack{a,d>0 \\ ad=n}} \sum_{b=0}^{d-1} d^{-k} F\left(\frac{az + b}{d}\right)$$

and on the Fourier expansion $F(z) = \sum a_n q^n$ ($q = e^{2\pi i z}$) we have explicitly

$$T_n \sum_{m=0}^{\infty} a_m q^m = \sum_{m=0}^{\infty} \left(\sum_{d|m,n} d^{k-1} a\left(\frac{nm}{d^2}\right)\right) q^m.$$

XII.9 One says that F is a *Hecke eigenform* if F is an eigenvector for all n; that is, if $T_n F = \lambda_n F$ for all n. This must be so if $k = 4, 6, 8, 10$ or 14 in the case of a linear transformation on $M(\Gamma, k)$, which is then one-dimensional; also for $k = 12, 16, 18, 20, 22$, or 26 on the space of cusp forms $S(\Gamma, k)$. Then one has that

$$\lambda_n a_m = \sum_{d|n,m} d^{k-1} a\left(\frac{nm}{d^2}\right).$$

Thus a Hecke eigenform has $\lambda_n a_1 = a_n$ for all n, so necessarily $a_1 \neq 0$ and normalizing to $a_1 = 1$ entails $\lambda_n = a_n$ for all n. It follows that the coefficients a_n of a normalized Hecke eigenform must satisfy

$$\sum_{n=1}^{\infty} a_n n^{-s} = \prod_p (1 - a_p p^{-s} + p^{k-1-2s})^{-1}.$$

Conversely, a Dirichlet series with such an Euler product corresponds to a modular form which is a Hecke eigenform.

XII.10 Hecke discovered that for every k, $M(\Gamma, k)$ has a basis of Hecke eigenforms. The bare bones of the argument are quite elementary. The Eisenstein series G_k is an eigenform because its corresponding Dirichlet series is a constant multiple of $\zeta(s)\zeta(s+1-k)$ and has an appropriate Euler product. Since $M(\Gamma, k)$ is the direct sum of the eigenspace generated by G_k and $S(\Gamma, k)$ it suffices to show that the space of cusp forms has a basis of eigenforms. However, the space $S(\Gamma, k)$ has an inner product

$$\langle F, G \rangle = \iint_{\mathcal{H}/\Gamma} y^k F(z)\overline{G(z)} \frac{dx\,dy}{y^2}$$

with the integral being over a fundamental domain of Γ. An accessible* computation now shows that the T_n are Hermitian with respect to this inner product (the *Petersson* inner product). Since the T_n commute with each other, they can be simultaneously diagonalized. It follows, using the usual straightforward arguments for proving basic results in inner product spaces, that there is an orthogonal basis of eigenforms spanning $S(\Gamma, k)$. These arguments also fairly immediately yield the conclusion that if $S(\Gamma, k)$ has dimension d, then the coefficients of its eigenforms are real algebraic integers of degree at most d.

XII.11 My brief description above considers Hecke eigenforms for the group $\Gamma = \mathrm{PSL}(2, \mathbb{Z})$. For its subgroups $\Gamma_0(N)$ some substantial amendments are necessary, principally to adjust the formulas for the bad primes, those dividing the level N. But now there is an additional problem. Some modular forms of level N are so because they already are modular forms at a lower level, dividing N. However, the space of modular forms turns out to be a direct sum of those *old* forms and of the *newforms*, fair dinkum of level N. It turns out, as shown by Atkin and Lehner, that one should and can concern oneself just with the subspace of newforms. There is a convenient summary of these things at pp. 15–18 in H. P. F. Swinnerton-Dyer and B. J. Birch, 'Elliptic curves and modular functions', Antwerp IV, *Lecture Notes in Mathematics* 476 (New York: Springer-Verlag, 1975), pp. 3–32.

XII.12 The adjective "dinkum" — "genuine" — is defined in Lecture VII. Its first formal mathematical use arose in the present context. Hendrik

*By that I mean that an unsophisticated reader can take the definitions and mindlessly carry the argument through. Indeed, unless one acts fairly mechanically, the situation will seem too frightening to do anything at all.

Lenstra was lecturing on CM at a meeting of the Australian Mathematical Society and started to speak about an α that was *really* ... complex; and was rudely interrupted by my shouting the suggestion "dinkum complex" from the back of the hall.

XII.13 "Euler, Lagrange, Legendre, and others studied transformations of the elliptic integrals

$$\int \frac{dx}{\sqrt{(1-x^2)(1-k^2x^2)}}$$

and they discovered that certain values of k, called *singular moduli*, gave elliptic integrals that could be transformed into complex multiples of themselves. This phenomenon came to be called *complex multiplication*." For example, the duplication of the lemniscate, briefly mentioned in the Notes following Lecture VII, is a case of twice multiplying by $\sqrt{-2}$. Our viewpoint, whereby complex multiplication is just a rather unusual way of finding a sublattice, rather trivializes this history.

XII.14 "[Niels] Abel observed that singular moduli and the roots of the corresponding transformation equations have remarkable algebraic properties." Namely these numbers generate *abelian* extensions of imaginary quadratic number fields, that is, extensions with abelian — commutative — galois group.* These notions were subsequently taken up by Kronecker. Kronecker had conjectured that every abelian extension of $\mathbb{Q}$ lies in some cyclotomic extension $\mathbb{Q}(e^{2i\pi/n})$; this became the Kronecker-Weber theorem. So abelian extensions of $\mathbb{Q}$ naturally arise from division values of the exponential function. Kronecker could now dream his *Jugendtraum* according to which *all* abelian extensions of imaginary quadratic number fields arise from values in the manner that Abel had observed.

XII.15 The foolish asides are mine. I have taken the wise remarks above from David Cox, *Primes of the Form* x^2+ny^2, (New York: Wiley-Interscience, 1989). A telegraphic review in the *American Mathematics Monthly* (December 1990) said of this book:

> A unique and sensational book. Its goal is to answer the question: Given a positive integer n, which primes have the form $x^2 + ny^2$ where x and y are integers? The complete answer leads the reader on a wonderful excursion through the core of classical and modern algebraic number theory — quadratic forms, reciprocity laws, class field theory, elliptic and modular functions. There are many rich exercises, and the text is loaded with interesting, historical remarks.

XII.16 The lattice $\alpha\tau\mathbb{Z} + \alpha\mathbb{Z}$ is $\mathrm{N}\,\alpha = |\alpha\overline{\alpha}|$ times as coarse as the original lattice. Here $\overline{\alpha}$ is the conjugate of α. For example, if $\alpha = \frac{1}{2}(\sqrt{-7}+1)$, then the index of the lattice $\alpha\Omega$, in the lattice $\Omega = \alpha\mathbb{Z} + \mathbb{Z}$, is 2.

---

*Let's pause at this remark to admire the anachronisms.

XII.17 For our purposes an *order* of a quadratic number field is a subset of its domain of integers which includes 1 and is closed under addition and multiplication. For example, the domain of integers of $\mathbb{Q}(\sqrt{-3}\,)$ is $\mathbb{Z}+\rho\mathbb{Z}$, where $\rho = \frac{1}{2}(\sqrt{-3}+1)$ is a cube root of unity. This is the *maximal* order. It happens that the orders $\mathbb{Z}+2\rho\mathbb{Z} = \mathbb{Z}+\sqrt{-3}\,\mathbb{Z}$ and $\mathbb{Z}+3\rho\mathbb{Z}$ also have unique factorization. The other two extra cases are the orders $\mathbb{Z}+2\sqrt{-1}\,\mathbb{Z}$ and $\mathbb{Z}+\sqrt{-7}\,\mathbb{Z}$. The last is "extra" because, of course, the domain of all integers of $\mathbb{Q}(\sqrt{-7}\,)$ is generated by 1 and $\frac{1}{2}(\sqrt{-7}+1)$.

It is, for example, a mildly instructive exercise to convince oneself, noting that the domain $\{a+ib : a,b \in \mathbb{Z}\}$ of Gaussian integers has unique factorization, that the same is true for the order consisting just of those elements of the shape $a+2ib$. But one does not have unique factorization for orders $\{a+fib : a,b \in \mathbb{Z}\}$ of *conductor* f different from 1 or 2.

On the other hand, take $\alpha = \frac{1}{2}(\sqrt{-163}+1)$ and $\beta = \sqrt{-163}$. Notice that $\alpha\overline{\alpha} = 41$. The remarkable fact that the numbers n^2+n+41 are prime for all integers n with $-40 \le n < 40$ is equivalent to the fact that the domain $\mathbb{Z}[\alpha]$ has unique factorization.* However, $164 = 2^2 \cdot 41 = (1+\beta)(1+\overline{\beta})$ displays two different factorizations of 164 into primes of the order $\mathbb{Z}[\beta]$.

XII.18 The observation concerning prime-producing polynomials is credited to Rabinowitz, concerning whom the following story is told by Mordell in his 'Reminiscences of an octogenarian mathematician' [*Amer. Math. Monthly* **78** (1971), pp. 952-961], at p. 959:

> In 1923, I attended a meeting of the American Mathematical Society held at Vassar College in New York State. Some one called Rainich from the University of Michigan at Ann Arbor, gave a talk upon the class number of quadratic fields, a subject in which I was then very much interested. I noticed that he made no reference to a rather pretty paper written by Rabinowitz† from Odessa and published in Crelle's journal. I commented upon this. He blushed and stammered and said, "I am Rabinowitz". He had moved to the USA and changed his name ... .

*The former fact is the canonical example to convince one to be careful with guessing by "induction". The "argument" traditionally goes: Set $P(n) = n^2+n+41$. Then $P(0) = 41$ is prime, $P(1) = 43$ is prime, $P(2) = 47$ is prime, $P(3) = 53$ is prime, $P(4) = 61$ is prime, ... Let's check a ' "random" case. Yes, $P(30) = 971$ is prime! Therefore $P(n) = n^2+n+41$ is prime for all n. ...

†Transliteration is, naturally enough, language dependent. Thus "Rainich" writes as G. Rabinow*itsch* in *Crelle* ['Eindeutigkeit der Zerlegung in Primzahlfactoren in quadratischen Zahlkörpern', *J. für Math.* in **142** (1913), 153–164] but is referred to as "Rabinow*itz*" in the *Monthly*.

ANNALS OF MATHEMATICS

TABLE OF CONTENTS

SECOND SERIES, VOL. 141, NO. 3

May, 1995

ANMAAH

Collage made from the front and back covers

LECTURE XIII

If you open the correct books in their appropriate places, you'll see the following story. You'll be told to consider the number $N_{p,m}$ of points of a smooth algebraic variety V defined over the finite fields $\mathbb{F}_{p^m}$, for values $m = 1, 2, \ldots,$ and thence to develop the local zeta function

$$Z_V(u) = \exp\left(\sum_{m=1}^{\infty} \frac{N_{p,m}}{m} u^m \right).$$

You'll be interested, because an elliptic curve E is essentially the case where the "variety" — the set of zeros of some set of polynomial equations — is defined by a single cubic polynomial. If the author is polite, she'll acknowledge that this definition is surprising, but that it works. In fact it takes only a moment to see that the definition of the local function is the natural one. I'll sketch the idea below.

But first let me say a few words about zeta functions of a number field K. One defines the Dedekind zeta function $\zeta_K(s)$ by

$$\zeta_K(s) = \sum_{\mathfrak{a} \neq 0} \frac{1}{N(\mathfrak{a})^s} = \prod_{\mathfrak{p}} \left(1 - \frac{1}{N(\mathfrak{p})^s}\right)^{-1}.$$

The sum is over the absolute norm $N(\mathfrak{a})$ of the nonzero integral ideals of K — the number of residue classes modulo $\mathfrak{a}$; and the product is over the prime ideals of (the domain of integers of) K. It is manifest that the case $K = \mathbb{Q}$ yields the Riemann ζ-function. One can add a twist to this definition by introducing *characters* χ. A character of K is, fairly loosely speaking, a multiplication-preserving map of the ideals of K into the *unit circle* — the subgroup of $\mathbb{C}$ of elements of absolute value 1. More precisely, the value 0 is allowed as well, so the map is into the union of $\{0\}$ and the unit circle. That provides the so-called L-series

$$L(K, s, \chi) = \sum_{\mathfrak{a} \neq 0} \frac{\chi(\mathfrak{a})}{N(\mathfrak{a})^s} = \prod_{\mathfrak{p}} \left(1 - \frac{\chi(\mathfrak{p})}{N(\mathfrak{p})^s}\right)^{-1}.$$

A Dirichlet character χ is a special case for which $\chi(n) = \chi(n \mod m)$ with $\chi(n) = 0$ if and only if n and m have a nontrivial common factor. Dirichlet* constructed the corresponding L-functions in the course of his

*The L of L-function is sometimes alleged to be the L of Gustav Lejeune Dirichlet.

proof that an arithmetic progression $(r + nd)$ of common difference d contains its fair proportion $1/\phi(d)$ of the primes, of course provided that $\gcd(r, d) = 1$. The trivial character χ_0, the one that is 1 for all $n \neq 0$, again yields the Riemann ζ-function.

Back in the middle of the nineteenth century it could already be seen, or at least sensed, that the critical thing was the behavior of $\zeta_K(s)$ at $s = 1$. Indeed, one could count ideals cleverly and obtain a limit formula

$$\lim_{s=1}(s-1)\zeta_K(s) = \frac{2^{r_1}(2\pi)^{r_2}}{w\sqrt{|d|}}hR$$

at least for the case of abelian extensions — those with commutative galois group. Here $r_1 + 2r_2 = n$ is the degree $[K : \mathbb{Q}]$, with r_1 the number of real fields conjugate to K and r_2 the number of complex conjugate fields; w is the number of roots of unity in K; d is the discriminant of the field; and R, the *regulator*, measures the "size" of the group of units of the field. The quantity h is the *class number* of K, indicating its deviation from unique factorization. In the case of cyclotomic fields $K = \mathbb{Q}(\zeta_p)$ of p th roots of unity of interest to Kummer one obtains the *residue*

$$\frac{2^{(p-3)/2}\pi^{(p-1)/2}}{p^{p/2}}hR\,.$$

It is this sort of thing that Kummer analyses to get his results on regularity, Bernoulli numbers, and the like.

Let's return to algebraic varieties. A point α in the algebraic closure $\overline{\mathbb{F}}_p$ belongs to some subfield $\mathbb{F}_{p^d} \subset \overline{\mathbb{F}}_p$ of degree d over $\mathbb{F}_p$. Then α, α^p, α^{p^d}, ..., $\alpha^{p^{d-1}}$ are distinct and this set is called a *prime divisor* $\mathfrak{P}$ with $\deg\mathfrak{P} = d$. There is an analogy that leads one to define $\mathrm{N}\mathfrak{P} = p^{\deg\mathfrak{P}}$.

Let c_d be the number of prime *divisors* of E of degree d. Then plainly, the number of points $N_{p,s}$ of E in $\mathbb{F}_{p^s}$ is given by $N_{p,s} = \sum_{d|s} dc_d$. To see this notice that each $\mathfrak{P}$ of degree d provides d points of E in $\overline{\mathbb{F}}_p$. Set

$$Z_V(u) = \prod_{\mathfrak{P}} \frac{1}{1-u^{\deg\mathfrak{P}}} = \prod_{1}^{\infty}\left(\frac{1}{1-u^n}\right)^{c_n}.$$

Then

$$\frac{Z_V'}{Z_V} = \frac{1}{u}\sum_{n=1}^{\infty}\frac{c_n n u^n}{1-u^n} = \frac{1}{u}\sum_{m=1}^{\infty}\Bigl(\sum_{d|m} dc_d\Bigr)u^m = u^{-1}\sum_{m=1}^{\infty}N_{p,m}u^m\,.$$

Thus

$$Z_V(u) = \prod_{\mathfrak{P}}\frac{1}{1-u^{\deg\mathfrak{P}}} = \exp\Bigl(\sum_{m=1}^{\infty}\frac{N_{p,m}}{m}u^m\Bigr)$$

the point being that

$$Z_V(p^{-s}) = \prod_{\mathfrak{P}}\frac{1}{1-p^{-s\deg\mathfrak{P}}} = \prod_{\mathfrak{P}}\frac{1}{1-(\mathrm{N}\mathfrak{P})^{-s}}\,.$$

This is the sense in which Z_V provides the p-part of the zeta function of the variety V. However,

$$Z_V(u) = \frac{P(u)}{(1-u)(1-pu)}$$

with P a polynomial having integer coefficients, of degree twice the genus g — which is at most $(n-1)(n-2)$ — and with zeros of absolute value $n = \deg f$; here f is the smooth curve defining the variety V. So in the case of an elliptic curve E, where $n = 3$, we have $\deg P = 2$. Since $Z_E(0) = 1$ we may write

$$P(u) = (1-\alpha_p u)(1-\beta_p u).$$

Plainly, the whole story of the $N_{p,m}$ is told by the "root" α_p and its conjugate $\overline{\alpha}_p = \beta_p$. The other parts of Z_E have nothing to do with the particular variety V. If we put $u = p^{-s}$, as intended, then

$$\prod_p (1-p^{-s})^{-1} = \zeta(s) \quad \text{and} \quad \prod_p (1-p^{1-s})^{-1} = \zeta(s-1)$$

are just manifestations of the ordinary ζ-function (they count points on the affine line and the projective line, respectively). In any case, these factors have nothing to say about the elliptic curve E.

It's traditional to write a_p for the *trace* $\alpha_p + \beta_p$ of α_p, and it's known that the norm of α_p is p. So $P(u) = 1 - a_p u + pu^2$. A minor calculation will confirm that $a_p = p+1-N_{p,1}$. In any case, we now see that the information about the number of points on an elliptic curve E defined over the finite field $\mathbb{F}_p$ is contained in the local factor $L_p(E,s) = (1-a_p u + pu^2)^{-1}$.

Now imagine our beginning with an elliptic curve E defined over the rationals $\mathbb{Q}$. We'd like to define and collect local factors to form a global object. It will be called the L-function* $L(E,s)$ of the curve.

The strategy is to find a suitable model for E defined over $\mathbb{Z}$ (that is, an equation with integer coefficients defining E) and then to deem that equation to be defined modulo p. Such "deeming" is called *reducing* E with respect to p. Well, we dealt with such reductions above. The only special cases are when E has *bad* reduction, that is, when the reduction E_p is no longer elliptic but has become a *singular* cubic curve. The bad primes will be exactly those that divide the discriminant Δ of E.

If one checks the Notes to Lecture IX, one finds that in the bad case E_p can have at most a singularity at one point, say, the origin. The badness of it all is now distinguished according as we are in the split case: E_p has distinct tangents at $(0,0)$, in which case $P(u) = 1+u$; or we are in the nonsplit case: E_p seems to have no tangents — meaning that it has distinct tangents defined over some quadratic extension of $\mathbb{F}_p$ — and $P(u) = 1-u$; or there is a *cusp*: E_p has coinciding tangents and $P(u) = 1$. These are

*I once heard Swinnerton-Dyer explain that a function is called an L-function if there are lots of conjectures about it, preferably in regions where it is not yet proved to exist.

honest results of computation. In these singular cases $P(u) = 1 - a_p u$ is of degree 1, and one can check that, indeed, respectively, $N_{p,1} = p+1-1$, $p+1+1$, or $p+1$. Now we may write

$$L(E,s) = \prod_{p \text{ good}} (1 - a_p p^{-s} + p^{1-2s})^{-1} \prod_{p \text{ bad}} L_p(E,s) = \sum_{n=1}^{\infty} \frac{a_n}{n^s}$$

knowing what each factor means. It's not hard to check that $a_m \cdot a_n = a_{m \cdot n}$ whenever $\gcd(m,n) = 1$, so the arithmetic function $a(n) = a_n$ is multiplicative. The Dirichlet coefficients a_n are just the a_p, when $n = p$ is prime; and $a_1 = 1$. With some mild pain one can find formulas for the a_{p^k} in terms of a_p.

In Lecture VI, I mentioned Hasse's proof of "the Riemann Hypothesis" for elliptic curves, showing that α_p has absolute value $\sqrt{p}$. The upshot here is that essentially — one should also take account of the number of divisors of n — $a_n = O(\sqrt{n}\,)$, entailing that the L-series converges when the real part σ of s is more than $\frac{3}{2}$. Mind you, the interesting value of $L(E,s)$ turns out to be $L(E,1)$. Disregarding the fact that the function makes no sense at $s = 1$, we sort of have

$$L(E,1) \approx \prod_{p \text{ good}} \frac{p}{p - a_p + 1} = \prod_{p \text{ good}} \frac{p}{N_p}.$$

I've omitted the bad factors because when talking nonsense it's ludicrous to be too precise about it, and in any case, because the qualitative point is that plainly the larger the N_p, the more surely that $L(E,1) = 0$, which is the simplest part of the Birch–Swinnerton-Dyer Conjectures.

Indeed, it is reasonably natural to make the following observations. On the one hand, one can estimate the number $N(A)$ of rational points $P(x,y)$ on the curve of *height* less than A. By "height" one need mean no more than that there is a bound whereby both $|x|$, $|y| < A$; but see Lecture XV. It turns out to be not too hard to confirm that if the group $E(\mathbb{Q})$ of rational points has rank r, that is $E(\mathbb{Q}) \cong \mathbb{Z}^r \oplus T$ for some finite group T — recall Lecture VII — then, with some constant $C > 0$,

$$N(A) \sim C(\log A)^{r/2} \qquad \text{as } A \to \infty .$$

The underlying ideas and methods closely follow the analogous, and rather better understood, case of L-functions of number fields.

On the other hand, back in the early sixties Birch and Swinnerton-Dyer saw, and "confirmed" with extensive calculation, that we should have an asymptotic formula as follows, with some constant $C' > 0$ related to C:

$$\prod_{p<x} \frac{N_p}{p} \sim C'(\log x)^r \qquad \text{as } x \to \infty .$$

That leads to their extraordinary conjectures about $L(E,1)$. In particular they suggest that if the group $E(\mathbb{Q})$ is infinite, then $L(E,s)$ should have a zero of order m at $s = 1$ and that $m = r$. And stretching the analogy with number fields almost beyond endurance, they propose the "shape" of the Taylor coefficient of $(s-1)^r$. All this, mind you, at a point where the function might not exist.

If one cares to believe the BSD conjecture, which includes the belief that $L(E, s)$ can be continued to the left of the line $\sigma = \frac{3}{2}$, then, as in Zagier's example cited in my Notes to Lecture VI, all sorts of things can be explained and "proved". Clearly, it would be a pleasant thing if all elliptic curves were "good", at least to the extent of ensuring that the BSD conjecture makes some sense. The Modularity Conjecture does just that. It proposes that all elliptic curves are "good", pretty well in just the desired sense. Wiles proves a large swathe of the Modularity Conjecture.

Notes and Remarks

XIII.1 I spoke of opening "the correct books". One such correct book is Kenneth Ireland and Michael Rosen, *A classical introduction to modern number theory*, Graduate Texts in Mathematics **84** (New York: Springer-Verlag, 1982); see Chapters 10 and 11.

XIII.2 You may have noticed that the N_p of this Lecture is one more than the N_p of Lecture VI. I'm sorry about that (particularly as it gives away the extent to which I copied the ideas and style of Zagier). But it didn't seem useful in Lecture VI to make a fuss about the rational point at ∞, while here, of course, we recognize this as a perfectly good point.

XIII.3 I quote the class number formula rather wantonly, just to impress. But the underlying ideas are clear enough. On the one hand, one splits the sum $\sum N(\mathfrak{a})^{-s}$ into h sums, each sum over just the ideals $\mathfrak{a}$ of the one ideal *class*. It is then not too hard to show that each of these sums has the same residue L at $s = 1$, telling us that the residue of $\zeta_K(s)$ at $s = 1$ is of the shape hL, some L, as asserted. The trick now is to evaluate Dirichlet's constant L. One deals with say, the principal class, the next problem being to select representative *elements* for the ideals of that class. Finally, the residue of the sum, now over suitable elements, is evaluated by comparison with an integral. The analogue of that for the Riemann ζ-function is just

$$\frac{1}{s-1} = \int_1^\infty x^{-s}\,ds \approx \sum_{n=1}^{\infty} n^{-s},$$

making it plain that its residue at $s = 1$ is indeed 1.

XIII.4 All these things are done from first principles, à la Kummer, for the case of cyclotomic fields in Harold M. Edwards, *Fermat's Last Theorem:*

a Genetic Introduction to Algebraic Number Theory, Graduate Texts in Mathematics **50** (New York: Springer-Verlag, 1977), pp. 181–224.

XIII.5 Dirichlet's Unit Theorem asserts that the group of units of a number field $\mathbb{K}$ has nontorsion part a free group of rank $r_1 + r_2 - 1$, where $\mathbb{K}$ has r_1 real conjugate fields over $\mathbb{Q}$ and r_2 pairs of complex conjugate such fields; thus $r_1 + 2r_2 = r = [\mathbb{K} : \mathbb{Q}]$, the degree of the field. The torsion part consists just of those roots of unity that are contained in $\mathbb{Q}$; the number of such roots of unity is frequently denoted by $|w|$. The BSD conjectures draw an analogy between the group of units of a number field and the group of rational points on an elliptic curve.

XIII.6 We can make the multiplicative group of units an additive group by dealing with the logarithms of the (ordinary) absolute values of the generators rather than with the generators themselves. That turns the group into a lattice in $\mathbb{R}$. Naturally, one asks about the volume of the fundamental domain of that lattice. That volume is called the *regulator* R of the number field. It easily follows from Dirichlet's arguments that we do indeed get a lattice, so this volume is nonzero.

XIII.7 I have mentioned Hasse's Theorem several times. There's no need for me to leave it a complete mystery. If I neglect all details, then the ideas behind its proof are straightforward. The trick is to study morphisms of elliptic curves. Accordingly, one defines an *isogeny* between elliptic curves E_1 and E_2 to be a morphism $\phi : E_1 \to E_2$ which satisfies $\phi(O) = O$. A special case is the zero, or constant, isogeny $\phi(E) = \{O\}$. Two elliptic curves are *isogenous* if there is a nonconstant isogeny between them.

The word "morphism" is there to remind us that an isogeny must be structure preserving; thus $\phi(P + Q) = \phi(P) + \phi(Q)$, for points P and Q on E_1. Mildly surprisingly, preservation of the group structure is almost automatic for seemingly arbitrary maps of elliptic curves, given that O is sent to O. To see why that might be so, recall that the collection $E(\mathbb{C})$ of all complex points on an elliptic curve E is isomorphic to a quotient $\mathbb{C}/\Lambda$, for some lattice Λ. It's not too difficult to convince oneself that the only analytic maps $\phi : \mathbb{C}/\Lambda_1 \to \mathbb{C}/\Lambda_2$ between two such curves are the affine maps $\phi : z \mapsto \alpha z + \beta$; thus just $\phi : z \mapsto \alpha z$ if 0 is sent to 0. If a map is well defined on the lattices, there will plainly be severe restrictions on α.

XIII.8 Indeed, in general a map $\phi : \mathbb{C}/\Lambda \to \mathbb{C}/\Lambda : z \mapsto \alpha z$ is not well defined unless $\alpha \in \mathbb{Z}$. The exception is the case of complex multiplication when the τ in the upper half-plane defining the lattice is in an imaginary quadratic number field; then α may lie in the ring of integers of that field.

XIII.9 Let's stay with these maps of an elliptic curve into itself, thus with the *endomorphisms* of a curve E. With each endomorphism ϕ one can associate its *degree*, $\deg \phi$. According to one's taste or emphasis, $\deg(\phi)$ may be viewed as the number of points of $E(\mathbb{C})$ sent to O or, on taking (x, y) as a *generic* point on the curve, preferably as the degree of the field $\mathbb{C}(x, y)$ over the field $\mathbb{C}(\phi(x), \phi(y))$. Clearly "multiplication by m" has degree m^2. Indeed, in general, given endomorphisms ϕ and ψ, there are integers a, b, and c so that an endomorphism $n\phi + m\psi$ has degree $an^2 + bnm + cm^2$, a quadratic form in m and n. Since degree is nonnegative, necessarily $a \geq 0$, $c \geq 0$, and $b^2 - 4ac \leq 0$. Since the only endomorphism of degree 0 is the zero endomorphism, it follows by a completely elementary argument that an endomorphism ϕ satisfies a quadratic equation $\phi^2 - s\phi + t = 0$.

The point of these observations is that all this holds for elliptic curves defined over any field, including finite fields. Of course in the remarks and definitions above we need to replace, wherever it is mentioned, the field of complex numbers $\mathbb{C}$ by the algebraic closure of the base field.

XIII.10 So what? Well, consider the number of elements of the group $E(\mathbb{F}_p)$ on an elliptic curve $E : y^2 = x^3 + Ax + B$ defined over the finite field $\mathbb{F}_p$ of p elements. By taking p th powers of both sides of the equation, we get

$$y^{2p} = x^{3p} + A^p x^p + B^p \quad \text{so} \quad (y^p)^2 = (x^p)^3 + Ax^p + B\,.$$

So the *Frobenius* map $\phi_p : (x, y) \mapsto (x^p, y^p)$ is an endomorphism of E. Moreover, consider the map $\overline{\mathbb{F}}_p \to \overline{\mathbb{F}}_p : x \mapsto x^p - x$. Plainly, its kernel is precisely $\mathbb{F}_p$; as usual $\overline{\mathbb{F}}_p$ denotes the algebraic closure of $\mathbb{F}_p$. Thus the kernel of $\phi_p - 1$ acting on the group $E(\overline{\mathbb{F}}_p)$ is precisely the points of the group $E(\mathbb{F}_p)$. Since ϕ_p is $1:1$ it follows that $\deg(\phi_p - 1) = \#E(\mathbb{F}_p)$.

Obviously, $\deg(\phi_p) = p$. So there is an integer s such that

$$\deg(\phi - 1) = p - s + 1\,,$$

where $s^2 - 4p \cdot 1 \leq 0$. Thus $s \leq 2\sqrt{p}$, which is Hasse's theorem

$$\left|\#E(\mathbb{F}_p) - (p + 1)\right| \;\leq\; 2\sqrt{p}\,.$$

The bound for s immediately entails the "Riemann Hypothesis", whereby the "eigenvalues" α and β of the Frobenius — that is, such that

$$(\phi_p - \alpha)(\phi_p - \beta) = \phi_p^2 - s\phi_p + p = 0$$

— must both have absolute value $\sqrt{p}$.

XIII.11 Apropos of counting points accurately, there is a positive definite quadratic form, the absolute canonical height, $\hat{h}\colon E(\mathbb{Q}) \otimes \mathbb{R} \to \mathbb{R}$ with $\hat{h}(P) - \log\max\{|x|, |y|\}$ bounded. One counts points in an r-dimensional ellipsoid of diameter $\approx (\log A)^{\frac{1}{2}}$ and finds that the constant C to be

$$C = \frac{\pi^{\frac{1}{2}r}|T|}{(\frac{1}{2}r)!\,\sqrt{R}}\,,$$

where R, the *regulator*, is the determinant of the symmetric $r \times r$ matrix defining $\hat{h}$ with respect to some $\mathbb{Z}$-basis of $E(\mathbb{Q})/T$. Particular special cases are $R = 1$, if $r = 0$ and $R = \hat{h}(P_0)$, if $r = 1$ and P_0 is a generator of $E(\mathbb{Q})/T$. There will be more of this in Lectures XV and XVI.

Ken Ribet at the Boston meeting, August 1995*

Photograph by C. J. Mozzochi, Princeton

*Notice that Ken has just written my name, $a_\ell(f)$, on the blackboard.

LECTURE XIV

It seems a pleasant idea to take a bit of a break from ferocious, though beautiful formulas and to chat a bit about conjectures and the like. There was a time when a conjecture in mathematics was on a par with a well-established theory in science. Consider, for example, the Riemann Hypothesis, concerning the zeros of the Riemann ζ-function

$$\zeta(s) = \sum_{n=1}^{\infty} \frac{1}{n^s} \qquad \text{with} \quad s = \sigma + it\,.$$

It is continued to the left of the line $\sigma = 0$ by the functional equation

$$\zeta(s)\Gamma(\tfrac{1}{2}s)\pi^{-\frac{s}{2}} = \zeta(1-s)\Gamma(\tfrac{1}{2}(1-s))\pi^{\frac{s-1}{2}}\,.$$

The function $\zeta(s)$ has a simple pole at $s = 1$ with residue 1, meaning that its Laurent expansion at $s = 1$ looks like

$$\zeta(s) = 1 \cdot (s-1)^{-1} + c_0 + c_1(s-1) + c_1(s-1) + c_2(s-1)^2 + \cdots\,.$$

It has "trivial" zeros at $s = -2, -4, \ldots$ compensating for the poles of $\Gamma(s)$. The striking fact is symmetry about the line $s = \frac{1}{2} + it$, showing that zeros of $\zeta(s)$ in the *critical strip* $0 < \sigma < 1$ are equally distributed about the critical line.

Pure thought as well as extensive calculation leads one to guess that all the nontrivial zeros of $\zeta(s)$ have real part $\frac{1}{2}$. When I say "extensive", I do mean very extensive calculation. By 1986 the first 1,500,000,001 zeros in the upper half-plane had been shown to be on the critical line [J. van de Lune, H. J. J. te Riele and D. T. Winter, 'On the zeros of the Riemann zeta function in the critical strip, IV', *Math. Computation* **46** (1986), pp. 667–681] and regions, considered most likely to contradict the RH, much higher up the line have been checked. The trouble is, our calculation is such a small part of infinity*. Were this science, the RH would

*"Eternal life doesn't put earthly life in the shade because it's *longer* ... Think of a ball of steel as large as the world, a fly alighting on it once every million years. When the ball of steel is rubbed away by the friction, eternity will not even have begun." As said in David Lodge, *The Picturegoers* (New York: Penguin, 1993), p. 147. I mention this not just because it's apropos, but to be able to add that any person who has not read Lodge's marvelous satires on academic life: *Changing Places*, *How Far Can You Go?*, *Nice Work*, and *Paradise News* is much the poorer for it.

have been convincingly verified. But it is mathematics and one does not have to be berserk to fear that the RH may be false.

However, it's the ABC-conjecture that moved me to write about such questions. In 1985 Masser and Oesterlé formulated a conjecture on the basis of analogy with a proved result for function fields. The idea is to define the "radical" or "conductor"

$$N(n) = \prod_{p \mid n} p$$

of an integer n as the product of the distinct primes dividing n. Of course, always $N(n) \le n$ with equality if and only if n has no repeated prime factor, that is, if n is *squarefree.* The conjecture claims that *if A, B, and C are positive integers satisfying $A + B = C$, then, say*

$$C < (N(ABC))^2 .$$

I add the qualification "say" because technically the exponent 2 may be replaced by any $1 + \varepsilon > 1$, provided that we admit a finite number of exceptions depending only on ε.

Certainly, the exponent does seem always to be less than 2. The current records, and the corresponding exponent, are, respectively,

$$3^{10} \cdot 109 + 2 = 23^5 \qquad \text{requiring exponent } 1.62991$$
$$3^2 \cdot 5^6 \cdot 7^3 + 11^2 = 2^{21} \cdot 23 \qquad \text{requiring exponent } 1.62599$$
$$19 \cdot 1307 + 7 \cdot 29^2 \cdot 31^8 = 2^8 \cdot 3^{22} \cdot 5^4 \qquad \text{requiring exponent } 1.62349$$

due to Reyssat, de Weger, and to Browkin and Brzezinski, respectively.

If $x^n + y^n = z^n$, then since $N(x^n y^n z^n) = N(xyz)$, plainly

$$z^n > (N(x^n y^n z^n))^2$$

once n is 6; the relevance of this to Fermat's Last Theorem is obvious. Moreover, we can outdo Fermat. When all's said and done, Fermat might just as readily have remarked in his margin that no power higher than a square can be split into two such higher powers — more precisely, that with exponents r, s, and t all at least 3, the equation

$$x^r + y^s = z^t$$

has no solutions in positive integers x, y, and z with $\gcd(x, y, z) = 1$.

Much concerning this question has recently been studied by Darmon and Granville. If we drop the constraint that each of the exponents be greater than 2, then there are solutions. Indeed, it turns out that the cases

$$\frac{1}{r} + \frac{1}{s} + \frac{1}{t} < 1 \quad \text{and} \quad \frac{1}{r} + \frac{1}{s} + \frac{1}{t} \ge 1$$

are quite different. Of course, we know all about $r = s = t = 2$ when all *primitive* solutions, that is with $\gcd(x, y, z) = 1$, are given by

$$\left(\tfrac{1}{2}(m^2 - n^2)\right)^2 + (mn)^2 = \left(\tfrac{1}{2}(m^2 + n^2)\right)^2$$

with m, n relatively prime and both odd. It's relatively easy to use the Pythagorean triples, just mentioned, to obtain parametric families of solutions for the cases $(r,s,t) = (r,2,2)$ and thence for $(r,s,t) = (2,2,t)$. The formula

$$(m^4 - 8mn^3)^3 + (4m^3n + 4n^4)^3 = (m^6 + 20m^3n^3 - 8n^6)^2$$

with $\gcd(m,n) = 1$, displays infinitely many primitive solutions of the equation $x^3 + y^3 = z^2$. Similarly one obtains infinitely many solutions of $x^4 + y^3 = z^2$ from

$$x = m^6 + 15m^4n^2 - 45m^2n^4 - 27n^6$$

$$y = -m^8 + 28m^6n^2 + 42m^4n^4 + 252m^2n^6 - 81n^8$$

$$z = 4mn(3m^{10} - 11m^8n^2 + 198m^6n^4 + 594m^4n^6 - 297m^2n^8 + 729n^{10})$$

with $\gcd(m,n) = 1$, $m + n$ odd and $3 \nmid m$. Thus $m = 5$ and $n = 2$ yields

$$33\,397^4 + 2\,161\,839^3 = 3\,368\,602\,840^2 .$$

The case $(r,s,t) = (2,3,5)$ probably also falls into this category.* Euler dealt with $(r,s,t) = (3,3,3)$; it happens to have no solution.

On the other hand, the ABC-conjecture suggests that there are just finitely many primitive solutions when $1/r + 1/s + 1/t < 1$. We'll take the exponent $1 + \varepsilon = 1029/1025$ for the conjecture, noting† that it now reads that there is some constant $k > 0$ so that $A + B = C$ has no primitive solutions if

$$C \geq k(N(ABC))^{1029/1025}$$

Of course $x^r < z^t$ and $y^s < z^t$. Then

$$z^t < k(N(x^r y^s z^t))^{1+\varepsilon} < k(xyz)^{1+\varepsilon} < k\, z^{(1+\varepsilon)t(1/r+1/s+1/t)} .$$

But it's a cute fact that if $1/r + 1/s + 1/t < 1$, then $1/r + 1/s + 1/t \leq 41/42$. Hence we have $z < kz^{49/50}$, and so any primitive solution of $x^r + y^s = z^t$ has $z < k^{50}$. Hence all of x, y, and z are bounded. Thus indeed, the equation can have just finitely many primitive solutions.

It's not too hard to notice the solutions $13^2 + 7^3 = 2^9$, $2^7 + 17^3 = 71^2$, $2^5 + 7^2 = 3^4$ and $3^5 + 11^4 = 122^2$. I suppose one might include $1 + 2^3 = 3^2$, if only out of respect for history, for it provides the only known solution to Catalan's problem of finding all solutions to $z^t - y^s = 1$. We'll deem that $1 = 1^7$, say. One's computer will probably get tired before finding any solutions beyond these five.

There's a very loose principle, known in the trade as the *Law of Small Numbers*, which would lead one to declare that these are all the solutions.

---

*See the Notes to this Lecture.

†I don't much like pretending, like this, to have foresight when of course it's hindsight.

Surprisingly, it just isn't so. In preparing his problems and exercises for the Utrecht *Fermatdag*, Frits Beukers' tireless computer found four more solutions; Don Zagier's found a fifth. These "large" solutions are a severe blow for the respect that the "Law" of Small Numbers normally inspires. They are

$$\begin{aligned} 17^7 + 76\,271^3 &= 21\,063\,928^2, \\ 1\,414^3 + 2\,213\,459^2 &= 65^7, \\ 33^8 + 1\,549\,034^2 &= 15\,613^3, \\ 9\,262^3 + 15\,312\,283^2 &= 113^7, \\ 43^8 + 96\,222^3 &= 30\,042\,907^2. \end{aligned}$$

One should not dare suggest that the ten solutions mentioned are all the primitive solutions. Yet it's intriguing that these solutions do not contradict the GFC, the *Generalized Fermat Conjecture* with which I opened these remarks. We're probably far from proving anything like the ABC-conjecture for numbers. Yet if A, B and C are polynomials, say defined over $\mathbb{C}$, then it's a very particular case of a *theorem* of Dale Brownawell and David Masser that $\deg C \le \deg N(ABC)$.

Some conjectures are suggested both by structure and by the strong support of numerical evidence. If they were part of natural science, they'd surely be accepted fact. The Riemann Hypothesis is such a case. The ABC-conjecture is only a little more than wishful thinking. There is no uniform method for "translating" arguments that work for function fields into ones valid for the corresponding results for numbers. Yet, generally, results for function fields do have numerical analogues. It's hard to shake the feeling that, somehow, the two sets of arguments are analogous. If they really are, we don't yet understand just how.

However, it's not hard to exhibit phenomena for which functions and numbers seem to behave quite differently. For example, it's easy to decide whether a polynomial $f(X)$ has distinct zeros. One just uses the Euclidean algorithm to find the greatest common divisor $\gcd(f, f')$ of f and its derivative. Then f is squarefree if and only this gcd is a constant. But to decide whether some integer n is squarefree appears to be essentially as difficult as to factorize n.

In the mid-sixties Swinnerton-Dyer remarked* that for abelian varieties — elliptic curves are the simplest case — "very few number-theoretical theorems have yet been proved." He adds that "the most interesting results are conjectures, based on numerical computation of special cases, and the fact that one has no idea how to attack these conjectures suggests that there must be important theorems which have not yet even been stated."

*H. P. F. Swinnerton-Dyer, 'An application of computing to class field theory', in J. W. S. Cassels and A. Fröhlich eds., *Algebraic Number Theory* (New York: Academic Press, 1967), Chapter XII, pp. 280–291.

Viewed with hindsight, this is an evident understatement. Birch and Swinnerton-Dyer had the insight and courage to try to evaluate $L(E, 1)$ for elliptic curves E over $\mathbb{Q}$. Of course, they could do so only when the L-function was known to exist at 1; thus in the CM case and for elliptic curves that could be parametrized by modular functions, since for these Shimura had shown that the L-function has an analytic continuation.

Amusingly, Swinnerton-Dyer remarks that "few such [modular elliptic] curves appear in the literature ... ; we have not been able to mechanize the process of finding such curves — though they appear to be extremely common ... ." That was to change within months, when Weil provided necessary and sufficient conditions for a Dirichlet series to be the Mellin transform of a cusp form for $\Gamma_0(N)$. In the light of the conjectures of Birch and Swinnerton-Dyer and of Tate, one now guessed that all elliptic curves defined over $\mathbb{Q}$ are modular.

Notes and Remarks

XIV.1 My remarks on the ABC-conjecture and on the idea of Darmon and Granville are greatly helped by the expositions of Rob Tijdeman, 'De Fermatvergelijking en enkele generalisaties', and of Frits Beukers, 'Oefeningen rond Fermat', at the *Fermatdag*, Utrecht, November 1993.

XIV.2 The "evidence in favor of" the RH that I find the most compelling concerns the *Hawkins primes*. Everyone knows the Sieve of Erastosthenes, whereby one obtains the primes by listing all the integers; declaring 2 a prime and thence striking out every second integer; discovering that 3 is unstruck; declaring it a prime and therefore striking out every third integer; discovering that 5 is unstruck; and so on.

1 2 3 ~~4~~ 5 ~~6~~ 7 ~~8~~ 9 ~~10~~ 11 ~~12~~ 13 ~~14~~ 15 ~~16~~ 17 ...
1 2 3 ~~4~~ 5 ~~6~~ 7 ~~8~~ ~~9~~ ~~10~~ 11 ~~12~~ 13 ~~14~~ ~~15~~ ~~16~~ 17 ...

Hawkins suggests a more random approach. On discovering that 2 is unstruck, one strikes out each succeeding number with probability $\frac{1}{2}$. If now m is the first survivor, one strikes out each succeeding number with probability $1/m$, and so on. The consequence is that the Hawkins *random sieve* produces lots of sets of Hawkins primes, which, statistically so to speak, behave just as do the genuine primes.

Now the RH, which refers to the honest-to-goodness primes because of the Euler product

$$\zeta(s) = \sum_{n=1}^{\infty} \frac{1}{n^s} = \prod_p \left(1 - \frac{1}{p^s}\right)^{-1}$$

is equivalent to the exponent $\frac{1}{2}$ in the error term in

$$\sum_{p<x} 1 = \int_2^x \frac{dt}{\log t} + O(x^{\frac{1}{2}+\epsilon}).$$

If one now looks for the error term in counting Hawkins' primes one finds that, almost always, Hawkins' primes obey the RH. Thus the Riemann Hypothesis is true if our primes are a "typical" set of Hawkins primes.

The source of the Hawkins random sieve is D. Hawkins, 'The random sieve' [*Math. Mag.* **31** (1958), pp. 1–3]. A more recent paper giving additional references is C. C. Heyde, 'A log log improvement to the Riemann Hypothesis for the Hawkins random sieve' [*Ann. Probability* **6** (1978), pp. 870–875]. I first heard of Hawkins' primes at a talk of Chris Heyde's and, since I found it in my files, apparently got him to send me his paper.

XIV.3 Erastosthenes leaves the unfortunate impression that 1 is a prime. It is not; but in a certain sense that is a fairly recent decision, motivated by considerations such as that $\mathbb{Z}/(p)$ is a field if p is prime, whereas $\mathbb{Z}/(1) = \{0\}$ is not. The "decision" is recent enough to cause the list of D. N. Lehmer, of 133 pages each of 5000 primes, to actually contain just the first 664999 primes [D. N. Lehmer, *List of Prime Numbers from* 1 *to* 10006721, Publ. 165, (Washington D.C.: Carnegie Institution of Washington, 1914), xvi+133pp.; 'A correction in the list of primes', *Bull. Amer. Math. Soc.* **32**, p. 902].

XIV.4 "But it's a cute fact that if $1/r + 1/s + 1/t < 1$, then

$$1/r + 1/s + 1/t \leq 41/42 \,."$$

A fact is *cute* if quite surprising on first sight but obvious on actually doing some trivial calculations. Do them.

XIV.5 Henri Darmon and Andrew Granville in fact discuss a somewhat more general equation

$$Kx^r + Ly^s = Mz^t$$

with integer coefficients K, L, and M and $\gcd(Kx, Ly, Mz) = 1$. But much more to the point, their result for $x^r + y^s = z^t$ is not based on the unproved and inaccessible ABC-conjecture but on the Mordell Conjecture, proved by Faltings in 1983. Thus it is a *theorem* that there are at most finitely many exceptions to the Generalized Fermat Conjecture. But the proof of the Mordell Conjecture is *ineffective*. We have no actual bound for the largest solution. Thus computation can only provide experimental evidence and cannot yield a proof of the GFC.

[March 1994] I wrote the present lecture and these notes purely on the basis of rumors of the preprint: Henri Darmon and Andrew Granville, 'On the equation $z^m = F(x, y)$ and $Ax^p + By^q = Cz^r$' University of Georgia Mathematics Preprint Series **28** Vol. **II** (1994). The paper is filled with all sorts of wonderful facts and arguments, of course well beyond (but also not including all) the facts I mention. Find the published version of the paper (it is to appear in *Bull. London Math. Soc.*) and read it.

XIV.6 In June 1994 (at a Computational Number Theory workshop in Rotterdam sponsored by the Stieltjes Institute) I was lucky enough to hear a truly beautiful lecture of Frits Beukers describing his investigations of the equation $x^r + y^s = z^t$ in the case $1/r + 1/s + 1/t \geq 1$. Beukers reported Don Zagier having spent a few evenings listing all the parametric families of solutions other than for the case $(2, 3, 5)$, which defeated even him. Frits showed that he had assimilated Klein's *Icosahedron** and used its equations to show that every solution of $x^2 + y^3 = z^5$ does indeed belong to an infinite parametrized family, and that there are just finitely many such families.

XIV.7 One might think it excessively fussy in the GFC — the Generalized Fermat Conjecture — to keep on emphasizing that x, y, and z should be relatively prime. But this is of the essence, for without this restriction one can generate seemingly nontrivial solutions at will. For example, following Tijdeman, one notices that since $2 + 11 = 13$ it follows that

$$(2^{12} \cdot 11^{28} \cdot 13^{30})^3 + (2^7 \cdot 11^{17} \cdot 13^{18})^5 = (2^5 \cdot 11^{12} \cdot 13^{13})^7 .$$

But the common factor cannot be removed just like that, as it may be in $x^n + y^n = z^n$.

XIV.8 Elsewhere, in one of my papers, I write that W. D. Brownawell and D. W. Masser ['Vanishing sums in function fields', *Math. Proc. Camb. Phil. Soc.* **100** (1986), pp. 427–434] show that if $u_0 + u_1 + \cdots + u_n = 0$, then each solution has projective height bounded explicitly in terms of the genus g of the function field F and the number $m(v)$ of elements among the u_i which are units at the respective values v of F. If they are S-units, this bound is $\frac{1}{2}n(n-1)(|S| + 2g - 2)$.

XIV.9 A quantity is called an S-unit if it is a divisor of 1 (thus a *unit* proper) except perhaps for finitely many primes (technically, *places*) which belong to some previously nominated finite set S containing $|S|$ elements. S should always contain the archimedean places. It turns out that almost all theorems true about fair dinkum units are immediately evident for the larger groups of S-units, so this is a very useful way of generalizing one's thinking.

XIV.10 I have already mentioned that Dirichlet's Unit Theorem asserts that the group of units of a number field $\mathbb{K}$ has non-torsion part a free group of rank $r_1 + r_2 - 1$, where $\mathbb{K}$ has r_1 real conjugate fields over $\mathbb{Q}$ and r_2 pairs of complex conjugate such fields; thus $r_1 + 2r_2 = n = [\mathbb{K} : \mathbb{Q}]$, the degree of the field. The torsion part consists just of those roots of unity that happen to be contained in $\mathbb{Q}$. In this spirit the free part of a group of S-units has rank $|S| + r_1 + r_2 - 1$.

---

*Felix Klein, *The Icosahedron* (New York: Dover, 1956).

XIV.11 I also mentioned that we can turn the multiplicative group of units into an additive group by considering the logarithms of the archimedean absolute values of the generators, rather than dealing with the generators themselves. That turns the group into a lattice in $\mathbb{R}$. The lattice volume, thus of its fundamental domain, is the *regulator* R of the number field. It is nonzero.

On the other hand, it remains just a conjecture of Leopoldt that the same is true for the so-called p-adic regulator, the generalization of this notion for a group of S-units. For abelian extensions all is well. We learned in the sixties that this follows from Baker's Theorem on linear forms in logarithms. In general, Leopoldt's Conjecture remains just that — a conjecture. Our lack of knowledge seems to betray the relatively primitive state of transcendence theory rather than being a matter of inner truth.

XIV.12 The conjectures of Birch and Swinnerton-Dyer seem to belong to the category of ideas that creep up on one. This genre of hypotheses will be readily recognized by scientists. In the case of the BSD conjecture there were results in the CM case which suggested analogies with the number field case. Then extensive computation showed there really did seem to be something to this resonance and analogy.

Once it became recognized that modular elliptic curves — also called "Weil curves" — were "good", one could hope that all rational elliptic curves are "good". That hope led to more favorable data appearing until, within a few decades, an unlikely thought seemed to have become a certainty.

XIV.13 Writing a little later than Swinnerton-Dyer in my closing remarks, Bryan Birch states that the definitive statement of the Birch–Swinnerton-Dyer Conjectures appears in H. P. F. Swinnerton-Dyer, 'The conjectures of Birch and Swinnerton-Dyer and of Tate' in T. A. Springer ed. *Proceedings of a Conference on Local Fields*, Driebergen 1966 (New York: Springer-Verlag, 1967), pp. 132–157.

XIV.14 Weil's paper is 'Über die Bestimmung Dirichletscher Reihen durch Funktionalgleichungen' [*Math. Ann.* **168** (1967), pp. 149–156]. It ends with the question whether perhaps every elliptic curve over $\mathbb{Q}$ satisfies the conditions under discussion. Weil remarks* that this "problematisch scheint" but "mag dem interessierten Leser als Übungsaufgabe empfohlen werden."

*that this seems questionable but is recommended to the interested reader as an exercise.

LECTURE XV

The questions for this Lecture are, very loosely: How big is a point? How high is a number? As an example, in just what sense is $355/113$ more complicated than $22/7$? Here it seems obvious that we might say that the *height* of a rational number a/b is $\max(|a|, |b|)$. Mind you, this is an imperfect definition; surely $2/4$ is no more complicated than $1/2$?

My reason for raising this topic at all is that we'll need some notion of the size of points on elliptic curves. In particular, with such a concept one can readily sketch a proof of the Mordell–Weil Theorem that the group $E(\mathbb{Q})$ of rational points on an elliptic curve E is indeed finitely generated. Recall that this means that there are finitely many points on E so that, starting with just them, the chord and tangent process yields all the rational points on E. The idea of that proof is this: One first notices that for any $m \geq 2$ the group $E(\mathbb{Q})/mE(\mathbb{Q})$ of rational points of E modulo the m-division points is finite. Remember, a point P on the curve is an m-division point if $mP = O$ is the distinguished point at infinity on the curve. Let $P_1, P_2, \ldots, P_s$ be a set of representatives of the points of E modulo the m-division points. Then for each rational point Q there is some P_i so that $Q - P_i = mS$, where S is some rational point on E. This is where we need an idea of "height". With a good notion of height one sees that the height of S is much less than the height of Q. This descent is eventually absurd because rational points on E cannot have arbitrarily small height. It follows that the Mordell–Weil group of rational points on E is generated by the P_i and perhaps some additional finite set of points Q of small height. The (minimal) number of free generators so obtained — that is, other than generators of finite order — is the rank of the group.

To me, the most interesting aspect of this argument is the notion of height itself, so I'll first chat a little about the easier case of height of algebraic numbers. It used not to be at all obvious how one should measure the size of an *algebraic* number α. Recall, a number is algebraic if it is the root of a polynomial equation with integer coefficients. The old-fashioned way was to take the defining polynomial $P_\alpha \in \mathbb{Z}[X]$ of α:

$$P_\alpha(X) = p_0X^r + p_1X^{r-1} + \cdots + p_r$$

to be irreducible with relatively prime coefficients and $P_\alpha(\alpha) = 0$; then α was said to have height $\max |p_i|$. For the case of a rational number a/b

presented in lowest terms that fits in, because it has defining polynomial $bX-a$. Other measures of height included the typist's nightmare "house": $\lceil\!\overline{\alpha}\rceil = \max_\sigma\{|\sigma\alpha|\}$, with the max taken over the conjugates $\sigma\alpha$ of α; a "denominator" d so that $d\alpha$ is an algebraic integer; and, of course, the degree r of α. Then there was the *length* $L(\alpha) = |p_0| + \cdots + |p_r|$ and the ... . And there were lots of lemmata comparing and relating the various measures so attributed to α.

As it happens, the absolute height, now generally agreed with, appears without the present normalization and in heavy disguise, in work of Mahler, where it is used to compute inequalities for the old-fashioned height. The Mahler measure $M(P)$ of a polynomial P is

$$M(P) = \exp\Big(\int_0^1 \log|P(e^{2\pi it})|dt\Big).$$

In effect by Jensen's theorem, one has (taking $P = P_\alpha$ as above) that

$$M(P_\alpha) = |p_0|\prod_\sigma \max\{1, |\sigma\alpha|\}.$$

One now sets $M(P_\alpha) = (H(\alpha))^r$. Since $|p_0|$ may be viewed as the product of those nonarchimedean values of the zeros of P_α that lie outside the unit circle, the height $H(\alpha)$ is precisely given by

$$r\log H(\alpha) = \log|a_0| + \sum_\sigma \max(0, \log|\sigma\alpha|) = \sum_v r_v \log^+|\alpha|_v,$$

with, as usual, $\log^+(\ ) = \max\{0, \log(\ )\}$ and the sum being taken over the appropriately normalized absolute values $|\ |_v$ of the field $\mathbb{K} = \mathbb{Q}(\alpha)$. Here r_v is the local degree, the degree of the complete field $\mathbb{Q}(\alpha)_v$ over the complete field $\mathbb{Q}_v$.

So the right definition comes to this: Given an algebraic number field $\mathbb{K}$, one defines the absolute logarithmic height $h(\alpha)$ of $\alpha \in \mathbb{K}$ by normalizing the absolute values $|\ |_v$ of $\mathbb{K}$ so that their restriction to the rational field $\mathbb{Q}$ coincide with the usual p-adic or real absolute values. Finally, one sets

$$h(\alpha) = ([\mathbb{K}:\mathbb{Q}])^{-1}\sum_v r_v \log^+|\alpha|_v.$$

If one prefers an honest height rather than its logarithm, as I do[†], set $H(\alpha) = \exp h(\alpha)$, as above. Then for a rational number a/b expressed in lowest terms, one has $H(a/b) = \max\{|a|, |b|\}$, just as with the old-fashioned height. The normalization by $[\mathbb{K}:\mathbb{Q}] = r$ is important so that the height of α is not affected by our replacing $\mathbb{K}$ by some extension field,

---

[†]While presenting a seminar in Paris in 1987 I got involved in an animated exchange[‡]with Serge Lang on whether I should follow Bombieri or French practice on this point; the capitalization was the compromise reached.

[‡]I nearly wrote "slanging match", but that would have been a grossly unreasonable exaggeration and, worse, a poor pun.

whence the name *absolute height*. It is fundamental that if $\alpha \neq 0$, one has the *product formula*

$$\sum_{v} r_v \log|\alpha|_v = 0 .$$

It is also convenient to define $h(0) = 0$. It is a theorem of Kronecker, and not completely trivial, that $h(\alpha) = 0$ if and only if α is a root of unity or zero. We also have the useful facts, which I may use below without further warning, that

$$\begin{aligned} h(1/\alpha) &= h(\alpha), \\ h(\alpha_1 + \cdots + \alpha_n) &\leqq h(\alpha_1) + \cdots + h(\alpha_n) + \log n, \\ h(\alpha\beta) &\leqq h(\alpha) + h(\beta) . \end{aligned}$$

By proper use of the absolute height, one deals with algebraic numbers in almost the same comfort as one deals with rational integers. For example, consider the well-known *fundamental lemma* of transcendence theory

$$\textit{If } n \in \mathbb{Z} \textit{ and } |n| < 1, \textit{ then } n = 0 .$$

By the product formula, one easily obtains its generalization to arbitrary places of a number field $\mathbb{K}$. The *fundamental inequality* of diophantine approximation states that

$$\alpha = 0 \quad \text{or} \quad \log|\alpha|_v \geq -rh(\alpha).$$

If $p/q \in \mathbb{Q}$ and $\alpha \neq p/q$, we easily deduce a precise form of Liouville's Theorem,

$$|\alpha - p/q| \geqq (2H(\alpha)\max(|p|, |q|))^{-r} .$$

Indeed, set $\beta = \alpha - p/q$. Then $h(\beta) \leqq h(\alpha) + h(p/q) + \log 2$ and the result follows from $h(1/\beta) = h(\beta)$ and the fundamental inequality.

Liouville's Theorem provides a criterion for algebraicity. It therefore readily allows the construction of numbers that are *not* algebraic, that is, *transcendental* numbers. The canonical example is $\sum_{n\geq 0} 10^{-n!}$; it cannot be algebraic because it has too many good rational approximations.

The (projective) height $H(\beta)$ of an $(m+1)$-tuple $\beta = (\beta_0, \ldots, \beta_m)$ of elements of $\mathbb{K}$ is given by $\log H(\beta) = ([\mathbb{K} : \mathbb{Q}])^{-1} \sum_v \max_{1\leq i\leq m} \log|\beta_i|_v$. In alluding to the height of a point $\beta = (\beta_0 : \beta_1 : \cdots : \beta_m)$ in the projective space $P^m(\mathbb{K})$, one uses the same notation.

One might well wonder what makes a good notion of height. I have a strong urge to mutter that one's stomach knows inner truth when it meets it, but more prosaically it seems that certain invariance properties and a preservation of multiplication are the thing. It's also particularly important that there be only finitely many relevant objects of height less than any given bound. In any case, one can make do with taking the height $H(P)$ of a point $P(x, y)$ on an elliptic curve just to be the mildly sophisticated height for points described immediately above.

It turns out that in nearly all contexts one deals with the "small" height h, the logarithm of the height H. For numbers, multiplicativity of H now entails that $h(\alpha\beta) = h(\alpha) + h(\beta)$. That's fine; however, it's kind of upsetting that it is mildly painful to estimate $h(\alpha + \beta)$. Seemingly, addition is a rather more complicated operation that multiplication. It's something we should have realized from the first moment we learned about adding fractions.

Naturally, it is yet more complicated to see how the height of rational points on an elliptic curve behaves under the operations on that group. Our crude height is only approximately multiplicative. But with some effort one can show that the difference $h(2P) - 4h(P)$ is bounded above and below by constants, depending only on the given curve; the lower bound needs some care. This result suggests that $h(P)$ is approximately quadratic in the logarithmic height of the coordinates of P. That suggests next showing that $h(P_1 + P_2) + h(P_1 - P_2) - 2h(P_1) - 2h(P_2)$ is similarly bounded above and below, by quantities depending only on the curve. That done, one can now prove that the sequence $(2^{-2n}h(2^nP))$ is Cauchy. Its limit $\hat{h}(P)$ is called the *canonical height* on the curve. It has just the right properties. For example, since plainly $\hat{h}(O) = 0$, it follows from $\hat{h}(nP) = n^2\hat{h}(P)$ that the height of any torsion point must be zero. Since there are only finitely many P of bounded height, the converse — that $\hat{h}(P) = 0$ only if P is a torsion point on the elliptic curve — holds as well. The numerical analogue is that $h(\alpha) = 0$ if and only if α is a root of unity or $\alpha = 0$.

I've gone on a bit about heights principally to emphasize that this idea is all important in arithmetic geometry and diophantine analysis. In the case of Fermat's Last Theorem, well, we met heights in Lecture I when dealing with exponent 4. There we started with a putative rational point $P(x/y, w/y^2)$ on the elliptic curve $E : (w/y^2)^2 = 1 + (x/y)^4$. That led to a point $(c/d, y/d^2)$ on the elliptic curve $E' : (y/d^2)^2 = (c/d)^4 - 4$ and finally to the point $Q(u/v, c/v^2)$ on the original curve. I did say that Q has smaller height than P, but not that in fact $2Q = P$.

This example is important because it provides the motivating example for Mordell's observation that $E(\mathbb{Q})/2E(\mathbb{Q})$ is finite. We see the phenomenon of *isogeny*, to wit of a map $E \to E'$ of elliptic curves defined over the ground field $\mathbb{Q}$, inducing a homomorphism of the group $E(\mathbb{Q})$ of rational points of E to the group $E'(\mathbb{Q})$ of rational points of E'. Such a map must have as kernel some finite subgroup of $E(\mathbb{Q})$. Thus the kernel must consist of the m-division points for some appropriate m. Therefore, Mazur's proof (alluded to in Lecture VII) that the torsion subgroups of $E(\mathbb{Q})$ are severely restricted becomes a matter of describing the possible isogenies of elliptic curves over $\mathbb{Q}$.

In Mordell's proof one notices that there is a group homomorphism of $E(\mathbb{Q})$, in effect into an appropriate multiplicative group of algebraic numbers modulo its squares, and with kernel $2E(\mathbb{Q})$. Finally, one shows that the image of the map is finite, so $E(\mathbb{Q})/2E(\mathbb{Q})$ is finite.

Remarkably, the proof just alluded to is not *constructive* — in general it does not give an infallible procedure for determining $E(\mathbb{Q})/2E(\mathbb{Q})$ even in principle; in general, $E(\mathbb{Q})$ cannot be determined. The Tate-Shafarevitch group and thence the Selmer groups, of which one reads in Wiles' proof, measure the obstruction to the effectivity of the argument.

The truth of the Modularity Conjecture now alleviates this situation by allowing the application of additional analytic machinery. In particular, the Birch–Swinnerton-Dyer Conjectures then precisely predict the rank of the group $E(\mathbb{Q})$. It is popularly guessed that any rank can occur, but the largest case constructed thus far has $E(\mathbb{Q})$ with rank just 21.

Notes and Remarks

XV.1 If your primary reason for reading this lecture was to see a detailed but accessible proof of Mordell's Theorem (the Finite Basis Theorem), then run off and order a copy of *Lectures on Elliptic Curves* by J. W. S. Cassels, London Mathematical Society Student Text **24**, (Cambridge: Cambridge University Press, 1991).

Cassels unkindly follows his introductory remarks with an observation, from A. Bremner and J. W. S. Cassels ['On the equation $Y^2 = X(X^2 + p)$', *Math. Comp.* **42** (1984), pp. 257–264] to the effect that, for example, the group of rational points on the elliptic curve $Y^2 = X(X+887)$ is generated by the point $(0,0)$ of order 2 and the point of infinite order

$$\left(\frac{37\,5494\,5281\,2716\,2193\,1055\,0406\,9942\,0927\,9234\,6201}{6215\,9877\,7687\,1505\,4254\,6322\,0780\,6972\,3804\,4100},\right.$$
$$\left.\frac{256\,2562\,6789\,8926\,8093\,8877\,6834\,0455\,1308\,9648\,6691\,5320\,4356\,6034\,6478\,6949}{4900\,7802\,3219\,7875\,8895\,9802\,9339\,9592\,8925\,0960\,6161\,6470\,7799\,7926\,1000}\right).$$

Joke aside, this is yet another emphatic reminder that our understanding of elliptic curves must be informed by theory. One cannot hope to stumble onto the facts experimentally without very deliberate intent.

XV.2 A more leisurely proof of the Finite Basis Theorem, admittedly just for a special case but with every i carefully dotted by Joe Silverman (in any case, not a single ı has been noticed by me), is given in *Rational Points on Elliptic Curves* by Joseph H. Silverman and John Tate, Undergraduate Texts in Mathematics, (New York: Springer-Verlag 1992). This book is an expanded version of lectures given by Tate in the early sixties of which we all sequentially had "photocopies of ever decreasing legibility".

XV.3 Incidentally, the confirmation that also for elliptic curves defined over algebraic number fields there is a severe restriction on the torsion subgroup and thus on the possible isogenies over the ground field was provided by Loïc Merel early in 1994.

XV.4 My claim about the Mahler measure of a polynomial is easy but not obvious. First take $P(X) = X - re^{2\pi i\tau}$ linear. Because

$$\int_0^1 \log|e^{2\pi it} - re^{2\pi i\tau}|\,dt$$

plainly is independent of the real number τ, it follows that $M(X - \alpha) = M(X - |\alpha|)$. Since it is obvious that $M(PQ) = M(P)M(Q)$ for polynomials P and Q — that is, that M is multiplicative — we have

$$M(X - \alpha^n) = \prod_{k=0}^{n-1} M(X - \alpha e^{2k\pi i/n}) = M(X - |\alpha|)^n = M(X - \alpha)^n.$$

So $\log M(X - \alpha) = n^{-1}\log M(X - \alpha^n)$ is independent of n. Thus taking n arbitrarily large, we readily confirm that this must be zero if $|\alpha| < 1$ and must be $\log|\alpha|$ if $|\alpha| \geq 1$, as asserted.

XV.5 In the thirties D. H. Lehmer asked a still unsolved question about algebraic integers near the unit circle. It's not too difficult to prove. The argument goes back to Kronecker that if all the conjugates $\alpha = \alpha_1, \alpha_2, \ldots, \alpha_r$ of an algebraic integer α lie on the unit circle, then α is a root of unity. Is more true? Is there an $\epsilon > 0$ independent of the degree r so that

$$M(\alpha) = \prod \max(1, |\alpha_i|) > 1 + \epsilon$$

or α is a root of unity? Several years ago at a West Coast Number Theory meeting, David Boyd asked Dick Lehmer, in my hearing, why he had not conjectured one way or the other but had only posed a question. We were firmly told that one *conjectured* only if one had compelling numerical data supporting one's suggestion. Nowadays, many a question or remark is called a "conjecture" just on a whim. It wasn't so in the good old days.

XV.6 Lehmer's question arises in a paper D. H. Lehmer, 'Factorization of certain cyclotomic functions' [*Ann. Math.* **34** (1933), pp. 461–479]. Substantial contributions to the problem have been made by David Boyd [see for example his survey, David W. Boyd, 'Speculations concerning the range of Mahler's measure', *Canad. Math. Bull.* **24** (1981), pp. 453–469]. However, the best known inequality,

$$\log M(\alpha) \gg \left(\frac{\log\log\deg\alpha}{\log\deg\alpha}\right)^3,$$

is due to E. Dobrowolski ['On a question of Lehmer and the number of irreducible factors of a polynomial', *Acta Arith.*, **34** (1979), pp. 391–401]. Now I wouldn't care to stake everything on truth of the conjecture — Lehmer did admit that nowadays he would claim the answer to be "yes" — but it is a reasonable bet that the algebraic integer α with smallest

Mahler measure $M(\alpha) > 1$ is the positive zero $\alpha \simeq 1.17628\ldots$ of the polynomial

$$X^{10} + X^9 - X^7 - X^6 - X^5 - X^4 - X^3 + X + 1,$$

as had been suggested by Lehmer.

XV.7 My remark about studying

$$h(P_1 + P_2) + h(P_1 - P_2) - 2h(P_1) - 2h(P_2)$$

may seem totally out of the blue, but it is not. Nor is "quadratic form" a sophisticated concept. In elementary linear algebra we encounter inner products $\langle u, v\rangle$ of vectors — the dot product is the standard example — and then define the length, or norm, of a vector v by $|v|^2 = \langle v, v\rangle$. It is a quadratic form in the coordinates of v. Then, indeed,

$$|u + v|^2 + |u - v|^2 - 2|u|^2 - 2|v|^2 = 0.$$

So if something is "approximately" a quadratic form, then the combination mentioned is a natural object of study.

By the way, the interrelationship between norms and inner products allows some nice remarks to be made about "heights" on a vector space [see A. J. van der Poorten and R. C. Talent, 'A note on length and angle', *Nieuw Arch. Wisk. (4e ser.)* **11** (1992), pp. 19–25].

Actually, the vector space analogue is quite instructive in the present context. The decomposition theorem for endomorphisms of vector spaces — I mean in effect, the theorem that every square matrix over $\mathbb{C}$ has a Jordan Canonical Form — is quite complicated. But for inner product spaces diagonalization of hermitian matrices, thence of normal ones, is relatively straightforward. Yet all that has changed is that the vector space is supplied with a height — the norm, thence with an inner product.

XV.8 Following Bombieri, one defines the height $H(a)$ of a sequence (a_h) of elements a_h of a number field $\mathbb{K}$ by

$$\log H(a) = ([\mathbb{K} : \mathbb{Q}])^{-1} \limsup_{h\to\infty} h^{-1} \sum_{v} \max_{0\le i\le h} \log |a_i|_v$$

with the sum and normalizations as above. Such a definition implies the height of a sequence to be an invariant under multiplication by a nonzero element of $\mathbb{K}$. Plainly, by the product formula, sequences (a_h) and (ca_h) have the same height. When needing to attach a height to the sequence of coefficients of a power series $\sum a_h X^h \in \mathbb{K}[[X]]$, this definition is quite appropriate. Invariance under multiplication by nonzero algebraic constants is clearly desirable. Moreover, the nonarchimedean values progressively pick up the lowest common multiple of the denominators of $a_0, \ldots, a_h$. Hence Bombieri's height is a suitable arithmetic measure of the growth of the sequence. The geometric progression $(1, \alpha, \alpha^2, \ldots)$ has height $H(\alpha)$. The harmonic sequence $(1, \frac{1}{2}, \frac{1}{3}, \ldots)$ has height e.

XV.9 Cassels remarks in 'Diophantine equations with special reference to elliptic curves' [*J. London Math. Soc.* **41** (1966), pp. 193–291], that in "most of the cases actually investigated the rank [of $E(\mathbb{Q})$] turns out to be very small: 0, 1, or 2; and it had been widely conjectured that there is an upper bound for the rank .... This seems to me implausible because the theory makes it clear that an abelian variety can only have high rank if it is defined by equations with very large coefficients ...."

I had endeavored several times during the last twenty years to read Cassels' survey (it's a book, really). After having more or less completed this book, and after having learned more at the Boston University meeting on *Fermat's Theorem*, I happened to look at the survey again and read it with profit. It gives invaluable insight into the ideas and conjectures underlying the work of the past thirty years.

XV.10 The current story of elliptic curves of large rank over $\mathbb{Q}$ seems to be the following. No doubt this is well known to those who know it well but not so well known to those who do not know it. Mestre [J.-F. Mestre, *C. R. Acad. Sci. Paris* **313**, ser. 1 (1991), pp. 139–142; pp. 171–174; *ibid.*, **314** ser 1 (1992), pp. 453–455] constructed elliptic curves over $\mathbb{Q}[T]$ with $\mathbb{Q}[T]$-rank at least 11 (then later 12). The idea is to pick a six-tuple $a = (\alpha_1, \dots, \alpha_6)$ of integers and to set $h(X) = \prod_i (X - \alpha_i)$; then $g(X) = h(X+T)h(X-T) \in \mathbb{Q}[T](X)$. The point is that then there are polynomials $q(X)$, $r(X)$ in $\mathbb{Q}[T](X)$ with $\deg g = 6$ and $\deg r \leq 5$ such that $g = q^2 - r$.

It is now manifest that the curve $Y^2 = r(X)$ contains the twelve $Q[T]$-rational points $(\pm T + \alpha_i, g(\pm T + \alpha_i))$. If, moreover, the six-tuple A is so chosen that the leading coefficient, c_5 say, of the polynomial $r(X)$ vanishes — that is, if r is of degree at most four — then $Y^2 = r(X)$ is a model for an elliptic curve over $\mathbb{Q}[T]$. For appropriate choices $A \in \mathbb{Z}^6$ [for example $(-17, -16, 10, 11, 14, 17)$ is such a choice] one now takes one of the cited points as origin O for the group. It is then not too painful to show that the remaining eleven are independent $\mathbb{Q}[T]$-rational points.

Indeed, Nigel Smart points out to me that one can choose the α_i via specialization and computation of regulator matrices *so that* $c_5 = 0$; *and* it is not too painful .... Presumably at this stage one also selects A so as to facilitate actually finding $\mathbb{Q}$-independent points, as alluded to below.

Finally, one specializes $T \to t = t_1/t_2 \in \mathbb{Q}$, so obtaining curves $E_{A,t}$ which are good candidates for having high $\mathbb{Q}$-rank. Loosely speaking, choosing the pairs (t_1, t_2) is a matter of testing for lots of rational points modulo p on $E_{A,t}$, with the p running over a suitable quantity of good primes. This means hoping for the best when it comes to actually finding $\mathbb{Q}$-independent points on $E_{A,t}$.

In a series of papers, Koh-ichi Nagao [*Proc. Japan Acad. Ser. A*, **68** (1992), pp. 287–289; *ibid.*, **69** (1993), pp. 291–293; and (with T. Kouya), *ibid.*, **70** (1994), pp. 104–105] gives elliptic curves over $\mathbb{Q}$ of rank, respectively, at least 17, then 20, and finally 21, the current record. That these ranks are indeed attained is shown by explicitly displaying the relevant numbers of $\mathbb{Q}$-independent points. Some ingenious computational tricks are necessary to make feasible the search for suitable t_1/t_2.

One charm of the subject of elliptic curves is large numbers with some inner meaning. However, it's hard to feel much the wiser on being told that the choices $A = (95, 71, 66, 58, 13, 0)$ and $t = 619/195$ yield a curve $E_{A,t}$ which is $\mathbb{Q}$-isomorphic to the curve with minimal Weierstrass model

$$\begin{aligned} y^2 + xy &= x^3 - 431\,09298\,07663\,33677\,95836\,20958\,91166x \\ &\quad + 5156\,28355\,53666\,43659\,03565\,27998\,71176\,90939\,15330\,88196 \end{aligned}$$

and that on it one may display as many as 20 $\mathbb{Q}$-independent points.

Photograph by C. J. Mozzochi, Princeton

Fermat today:
Fermat unamused by a muse

La Salle des Illustres dans les galeries du Capitole de Toulouse

Photograph by Kate van der Poorten

LECTURE XVI

A mathematical proof should resemble a simple and clear-cut constellation, not a scattered cluster in the Milky Way. A chess problem also has unexpectedness, and a certain economy; it is essential that the moves should be surprising, and that every piece on the board should play its part.

G. H. Hardy, *A Mathematician's Apology**

Before turning to Wiles' proof, suppose that we look at a different classical problem solved by work of Dorian Goldfeld, and of Benedict Gross and Don Zagier on elliptic curves. The background is the remarks of Gauss on *quadratic forms.* The question concerns the collection of integers

$$\{n : n = Ux^2 + Vxy + Wy^2 \text{ for some pair of integers } (x, y) \neq (0, 0)\},$$

that is, the integers *represented* by a given quadratic form

$$Ux^2 + Vxy + Wy^2.$$

It's not useful for the integer coefficients U, V, and W to have a common factor, so we shall suppose not. That is, we suppose the form is *primitive.*

If a quadratic form $U'x'^2 + V'x'y' + W'y'^2$ is obtained from the form $Ux^2 + Vxy + Wy^2$ by a substitution $x' = ax + by$, $y' = cx + dy$, where $\begin{pmatrix} a & b \\ c & d \end{pmatrix}$ has integer entries and is of determinant ± 1, then the two forms represent exactly the same integers. One says that the forms are equivalent. It's easy to check that $V^2 - 4UW = V'^2 - 4U'W'$, so equivalent forms have the same *discriminant*. Clearly, equivalence is an equivalence relation on forms of given discriminant. Also *proper* equivalence, where the admissible transformations are restricted to determinant $+1$, is an equivalence relation on quadratic forms.

However, for example, the forms $x^2 + 5y^2$ and $2x^2 + 2xy + 3y^2$, though both of discriminant -20, are not equivalent. Specifically, the primes p represented by $x^2 + 5y^2$ are those such that $p \equiv 1$ or $9 \pmod{20}$; those represented by $2x^2 + 2xy + 3y^2$ are $\equiv 3$ or $7 \pmod{20}$. For the latter

*I was reminded of this quotation by its appearance in *Enigma*, a recent novel by Robert Harris (London: Hutchinson, 1995).

Fermat had noticed that, apparently, the product of any two primes $\equiv 3$ or $7 \pmod{20}$ *is* represented by $x^2 + 5y^2$. Why might this be so?

First let's notice that the number $h = h(\Delta)$ of inequivalent forms of given discriminant $\Delta < 0$ is finite. It is the same thing to say that the number of *equivalence classes* of forms of that discriminant is finite. All we need is to confirm that each equivalence class contains a form with $|V| \le U \le W$. Recalling that Δ is negative, it follows for reduced forms that $3U^2 \le -\Delta$, so there are only finitely many such *reduced* forms and hence certainly only finitely many equivalence classes. The case $\Delta < 0$ is that of positive definite forms. The analogous argument for indefinite forms is a little more complicated.

In any event, here I will want to concentrate on the definite case. One can check that, up to proper equivalence, there is precisely one reduced form in each equivalence class. Using this, it is not terribly difficult to compute the *class number* $h(\Delta)$ for any given $\Delta < 0$. Gauss did just that for many Δ and noticed that the sequence of discriminants with a given class number h seems to end for each value of h. Thus the last Δ with $h(\Delta) = 1$ is apparently -163, the last with $h = 2$ seems to be -427, and the last with $h = 3$, -907. The proof of this remained an entirely open problem for over a hundred years.*

Around 1916 Hecke found a lower bound for $h(\Delta)$ in terms of $|\Delta|$, subject to the Riemann Hypothesis generalized to its analogue for number fields; and in 1933 Deuring showed that the *falseness* of the ordinary Riemann Hypothesis entails $h(\Delta) > 1$ if $|\Delta|$ is large enough. Soon after, Mordell showed that indeed $h(\Delta) \to \infty$ if the RH is false, and in 1934 Heilbronn proved this if the generalized Riemann Hypothesis is false. So together with Hecke's result, one had an unconditional proof of the finiteness of the number of Δ for which $h(\Delta)$ has any given value. A year later Siegel made all this explicit, by showing that $h(\Delta) > C(\varepsilon)|\Delta|^{\frac{1}{2}-\varepsilon}$ as $-\Delta \to \infty$ for any $\varepsilon > 0$.

But like the earlier results, this bound was *ineffective*. To know that one had found all instances of any particular class number one had to either, find an additional unexpectedly large $|\Delta|$ with that small class number, or say, prove the GRH. So in practice† one actually was no nearer to knowing whether Gauss had found all discriminants with small class number.

Then in the mid-sixties the class number 1 problem was settled by Harold Stark and independently by Alan Baker. Stark's work suggested that clearer writing would have allowed Heegner (1952) to have claimed

---

*Here and below I selectively quote from Don Zagier's beautifully written '*L*-series of elliptic curves, the Birch-Swinnerton-Dyer Conjecture, and the class number problem of Gauss' [*Notices Amer. Math. Soc.* 31 (1984), pp. 739–743]. Forget this vulgarization and go and read it.

†In theory, there's no difference between theory and practice. In practice, however, it doesn't quite work that way.

that success. Baker and Stark also jointly settled the case $h = 2$, using Baker's method; that approach did not work for larger class numbers.

The final breakthrough came in 1975 with Dorian Goldfeld proving a deep theorem to the effect that the existence of a single L-function with a zero of sufficiently high order at the symmetry point of its functional equation could be used to give an effective lower bound for $h(\Delta)$ going to infinity as $\Delta \to -\infty$. Gross and Zagier produced such a function.

If conjectures were facts, there would have been next to nothing to this. The calculations initiated by Birch and Swinnerton-Dyer had long suggested that in the neighborhood of the symmetry point $s = 1$ the L-series of an elliptic curve E is

$$(s-1)^m \frac{R}{|T|^2} \Omega \,|\,Ш\,| .$$

Here $m = r_{\mathrm{an}}$, the order of vanishing of the L-function at 1, is called the analytic rank of the curve and is conjecturally the same as the (geometric) rank r. R is the regulator; if $m = 1$, this is just $\hat{h}(P)$, the absolute height of a generator P of the infinite part of $E(\mathbb{Q})$. $|T|$ is the order of the torsion part of $E(\mathbb{Q})$. Further, $\Omega > 0$ is a simple rational multiple (depending on the bad primes) of the real period of the elliptic curve, and $|\,Ш\,|$ is the order of the Tate–Shafarevitch group Ш (pronounced "sha"). More of that too later, but here it is worth remarking that in general the group Ш is not even known to be finite. Yet somehow this remaining factor was "recognized" as surely being the putative finite order of that group.

In any case, what Gross and Zagier did was to prove that if E is a Weil curve (thus, a "modular" elliptic curve), then $m = 1$ entails $r \geq 1$. That is, if $L(E,1) = 0$ and $L'(E,1) \neq 0$, then the group $E(\mathbb{Q})$ is infinite. Moreover, $L'(E,1)$ then is a rational multiple of ΩR, and this multiple can sometimes be shown to be a square — it is known that if Ш is finite, then its order is a square. Finally, they showed that there do exist Weil curves E with $m = r = 3$; for example, the curve $E : -139y^2 = x^3 + 10x^2 - 20x + 8$ of conductor $N = 37$. That provides the required L-function with a zero of order 3 at $s = 1$. The upshot is that

$$h(\Delta) > C \prod_{p|\Delta} \left(1 - \frac{2}{\sqrt{p}}\right) \cdot \log|\Delta|$$

for some effectively computable absolute constant $C > 0$. That suffices to settle Gauss's conjecture for positive definite quadratic forms.

In 1967 Weil had shown that the standard conjectures on the analytic continuation and functional equation of the L-series of an elliptic curve E, and of its twists by Dirichlet characters, entails the modularity of E.

Conversely, there is a finite algorithm allowing one to determine all modular elliptic curves of given conductor N. Details may be found in the introduction to J. E. Cremona, *Algorithms for Modular Elliptic Curves*

(Cambridge: Cambridge University Press, 1992). The algorithm determines all elliptic curves arising as quotients of the Jacobian of the modular curve $X_0(N)$. A table of modular elliptic curves with the first several hundred conductors appeared in Antwerp IV. More recently John Cremona's book provides a list up to $N = 2\,000$; further of his calculations — he has now gone beyond 5 000 — are available on the internet.†

For a modular elliptic curve the L-function $L(E,s)$ is entire, and we know the parity of its order m at $s = 1$. It is even or odd according as the sign in the functional equation satisfied by $L(E,s)$ is $+1$ or -1. In turn, that depends on whether the corresponding modular form f of weight 2 satisfies $f(-1/Nz) = \mp Nz^2 f(z)$ — notice the reversal of sign.

Suppose that $\Delta < 0$ is the discriminant of an imaginary quadratic number field $\mathbb{K}$, that $(\Delta, N) = 1$, and β is some integer so that $\Delta \equiv \beta \pmod{4N}$. Then the set of z in the upper half-plane $\mathcal{H}$ satisfying a quadratic equation $Az^2 + Bz + C = 0$ with $N \mid A$, $B \equiv \beta \pmod{2N}$, and C an integer such that $\Delta = B^2 - 4AC$ is readily seen to be $\Gamma_0(N)$-invariant. Moreover, there are just finitely many equivalence classes of these z modulo the action of $\Gamma_0(N)$. Indeed, the number of such classes is precisely h, the class number of $\mathbb{K}$. We may view representatives $z_1, z_2, \dots, z_h$ of these classes as elements of $X_0(N)$. If $E: y^2 = g(x)$ is a modular elliptic curve of conductor N, there is a map $\phi: X_0(N) \to E(\mathbb{C})$ and following ideas of Heegner one considers the sum $\phi(z_1) + \cdots + \phi(z_h) = P_\Delta$ on $E(\mathbb{C})$. In fact, while the individual $\phi(z_i)$ have coordinates in some algebraic extension of $\mathbb{K}$ — the Hilbert class field of $\mathbb{K}$ — their sum P_Δ is in $E(\mathbb{K})$. Moreover, under complex conjugation P_Δ goes to $-\varepsilon P_\Delta$, where ε is the sign in the functional equation for $L(E,s)$. If $\varepsilon = +1$, then $2P_\Delta$ has the form $(x, y\sqrt{\Delta})$ and so gives a rational point (x,y) on the *twisted* curve $E^{(\Delta)}: \Delta y^2 = g(x)$. If, on the other hand, $\varepsilon = -1$, then $L(E,1) = 0$; then by the BSD conjecture one expects E to have odd, and thus positive rank, and $2P_\Delta \in E(\mathbb{Q})$.

Gross and Zagier prove that if $\varepsilon = -1$, then

$$L(E^{(\Delta)}, 1)L'(E, 1) = c \cdot \Omega_{E^{(\Delta)}} \cdot \Omega_E \cdot h_E(2P_\Delta)\,,$$

and if $\varepsilon = +1$, then

$$L(E,1)L'(E^{(\Delta)}, 1) = c \cdot \Omega_{E^{(\Delta)}} \cdot \Omega_E \cdot h_{E^{(\Delta)}}(2P_\Delta)\,.$$

Here $\Omega_{E^{(\Delta)}}$ and Ω_E are the periods occurring in the BSD conjecture and h_E is the absolute height on E, while $h_{E^{(\Delta)}}(2P_\Delta)$ is that on $E^{(\Delta)}$; c is a simple rational number. Notice that if $m = 1$, then $\varepsilon = -1$ and therefore $L'(E,1) \neq 0$. A theorem of Waldspurger guarantees the existence of a Δ such that $L(E^{(\Delta)}, 1) \neq 0$. So $2P_\Delta$ has nonzero height and thus must have infinite order in $E(\mathbb{Q})$. So, indeed, $m = 1$ entails $r \geq 1$. Moreover, for a Weil curve E, then $L(E,1)/\Omega_E$ could be proved to be rational; so also $L'(E,1)/h_E(2P_\Delta)$ is rational. But if $r = 1$, then $h_E(2P_\Delta)$ is a square integral

† At `ftp://euclid.exeter.ac.uk/pub/cremona` .

multiple of the regulator R of E. The strength of the identities mentioned is the absence of the mysterious quantity Ш.

Now $E: y^2 = x^3 + 10x^2 - 20x + 8$ is a Weil curve with conductor $N = 37$ and rank 0 so $L(E,1) \neq 0$, while it happens that $P_{-139} = 0$. Thus, necessarily, $L'(E^{(-139)},1) = 0$. Since, moreover, it can be shown that the sign in the functional equation for $E^{(-139)}$ is -1 and that the third derivative $L'''(E^{(-139)},1)$ does not vanish, it follows that the elliptic curve $E^{(-139)}: -139y^2 = x^3 + 10x^2 - 20x + 8$ has $m = 3$; it's also easy to show that its geometric rank r is 3. This "showing" and "seeing" is largely a matter of computation since it is easy to verify numerically that some quantity cannot be 0. On the other hand, the identities are essential to show that various quantities vanish, since for numbers that are potentially transcendental or zero, vanishing cannot readily be verified by numerical computation alone. Whatever, $E^{(-139)}$ provides the L-function necessary, given Goldfeld's result, to prove Gauss's conjecture.

By now a few more bits and pieces of the BSD conjecture have become accessible. At the time of the Gross and Zagier work we only had the theorem of Coates and Wiles to the effect that, for E with complex multiplication, $L(E,1) = 0$ if $E(\mathbb{Q})$ is infinite. Then following work of Thaine in cyclotomic fields, Karl Rubin proved the first instances of a finite Tate–Shafarevitch group and, aided by seminal work of Kolyvagin, materially simplified and generalized the Mazur and Wiles proof* of the "main conjectures" of Iwasawa theory.

The methods of Kolyvagin, aided by results of Gross and Zagier, Bump, Friedberg and Hoffstein, and R. Murty and K. Murty, now provide us with an important part of the Birch–Swinnerton-Dyer Conjectures. If E is a modular elliptic curve and either $L(E,1) \neq 0$ or $L'(E,1) \neq 0$, then Ш $=$ Ш$(E/\mathbb{Q})$ is finite and $E(\mathbb{Q})$ is, respectively, finite or of rank 1, exactly as predicted by BSD.

Notes and Remarks

XVI.1 I wrote that Fermat had noticed that the product of any two primes represented by $2x^2 + 2xy + 3y^2$ is represented by $x^2 + 5y^2$. That's easy to see by the identity

$$(2x^2 + 2xy + 3y^2)(2x'^2 + 2x'y' + 3y'^2)$$
$$= (2xx' + xy' + x'y + 3yy')^2 + 5(xy' - xy')^2.$$

But what's going on here?

*There is the happy side effect of a theorem still referred to as the "Main Conjecture"; contrasting pleasantly with conjectures, like the FLT until recently, known as "theorems". In that spirit I have no doubt that if ever proved, the Riemann Hypothesis will remain known as the RH, and the generalized Riemann Hypothesis as the GRH.

XVI.2 The product $Ux^2 + Vxy + Wy^2$ and $U'x'^2 + V'x'y' + W'y'^2$ of two quadratic forms is a nasty expression

$$UU'x^2x'^2 + UV'x^2x'y' + UW'x^2y'^2 + VU'xx'^2y + \cdots \qquad \text{etc.}$$

Suppose that it happens to happen that there is a substitution

$$\begin{aligned} X &= Axx' + Bxy' + Cx'y + Dyy' \\ Y &= A'xx' + B'xy' + C'x'y + D'yy' \end{aligned}$$

whereby that product becomes $uX^2 + vXY + wY^2$. That is much more palatable and allows us to say learnedly that this last form is a *compound* of the two given forms.

Of course, it is not at all obvious that such a bilinear substitution will ever exist, particularly when we insist that all of $U, V, \ldots, A, B, \ldots$ are to be integers, and that $A, \ldots, A', \ldots$ not all share a common factor. Nonetheless, the obvious example

$$(x^2 + y^2)(x'^2 + y'^2) = X^2 + Y^2$$

with $X = xx' - yy'$ and $Y = xy' + x'y$, or the more interesting one of Lagrange cited above, shows that we can get lucky.

XVI.3 That suggests the opposite concern. Perhaps there is always a suitable bilinear substitution whereby any two forms compound to yield any form whatsoever. That's not so. Two forms may be compounded if and only if they have the same *fundamental discriminant*, that is, if their discriminants are the same up to square factors.

To see that, we first remark that the form $\Phi = Ux^2 + Vxy + Wy^2$ has discriminant $V^2 - 4WU$. After the linear substitution

$$x = aX + bY \qquad y = cX + dY$$

we have a form

$$\begin{pmatrix} X & Y \end{pmatrix} \begin{pmatrix} a & c \\ b & d \end{pmatrix} \begin{pmatrix} U & \frac{1}{2}V \\ \frac{1}{2}V & W \end{pmatrix} \begin{pmatrix} a & b \\ c & d \end{pmatrix} \begin{pmatrix} X \\ Y \end{pmatrix}$$

which evidently has discriminant

$$\begin{vmatrix} a & b \\ c & d \end{vmatrix}^2 (V^2 - 4WU).$$

The point is that the linear transformation multiplied the discriminant by a square. Now comes a cute remark. Our bilinear substitution may be viewed as

$$\begin{aligned} X &= (Ax' + By')x + (Cx' + Dy')y\,, \\ Y &= (A'x' + B'y')x + (C'x' + D'y')y\,, \end{aligned}$$

a linear transformation of x and y. Hence viewed as a form in x and y, we see that on the one hand, $\Phi(X,Y) = uX^2 + vXY + wY^2$ has discriminant $\square(V^2 - 4WU)$; on the other hand, $\Phi(X,Y) = \varphi(x,y)\varphi'(x',y')$, still viewed as a form in x and y, plainly has discriminant $\square'(v^2 - 4wu)$. Of course, $\square$ and $\square'$ denote unspecified squares. Thus the discriminant $v^2 - 4wu$ differs from $V^2 - 4WU$ by multiplication by a rational squared. By symmetry, or laborious repetition of the argument *mutatis mutandis*, the same is true for the discriminant $v'^2 - 4w'u'$ *vis à vis* $V^2 - 4WU$. Hence the discriminants $v^2 - 4wu$ and $v'^2 - 4w'u'$ differ from one another by multiplication by a rational squared. One can summarize all this by reporting that:

> If forms $U(x - \gamma y)(x - \overline{\gamma}y)$ and $U'(x' - \gamma' y')(x' - \overline{\gamma}' y')$ possess a compound form $u(X - \Gamma y)(X - \overline{\Gamma}Y)$, then all three elements γ, γ', and Γ generate the same quadratic number field over $\mathbb{Q}$.

XVI.4 The problem now is to show that two forms of the same fundamental discriminant can be compounded. We'll discover an explicit way of doing that; I'll call this particular method *composition.*

We'll need some careful notation. Roman letters will represent rational integers. I let δ denote an integer of some quadratic number field, and suppose that δ has norm $\delta\overline{\delta} = n$ and trace $\delta + \overline{\delta} = t$. I write elements $\alpha \in \mathbb{Q}(\delta)$ in *canonical form* $\alpha = (K\delta + P)/Q$ with

$$Q \mid (K\delta + P)(K\overline{\delta} + P) = K^2n + KtP + P^2 .$$

That can be done, for if the condition does not hold, there is a divisor L of Q so that $LQ \mid ((LK)^2n + (LK)t(LP) + (LP)^2)$, and we replace the various parameters by their respective products with L. The point is to ensure that the quadratic form

$$Q(x - \alpha y)(x - \overline{\alpha}y) = Qx^2 - (2P + Kt)xy + \frac{K^2n + KtP + P^2}{Q}y^2$$

has integer coefficients.

XVI.5 Set $\alpha = (K\delta + P)/Q$ and $\alpha' = (K'\delta + P')/Q'$. We may suppose that K and K' are relatively prime, for if not we just replace δ by some multiple of itself. We attempt to compose the forms

$$Q(x - \alpha y)(x - \overline{\alpha}y) \quad \text{and} \quad Q'(x' - \alpha' y')(x' - \overline{\alpha}' y')$$

by considering just

$$\begin{aligned} G(x - \alpha y)(x' - \alpha' y') &= Gxx' - G\alpha' xy' - G\alpha x' y + G\alpha\alpha' yy' \\ &= X - \beta Y \\ &= (A - \beta A')xx' + (B - \beta B')xy' + (C - \beta C')x'y + (D - \beta D')yy'. \end{aligned}$$

Given human limitation, it seems wise to simplify by taking $K = K' = 1$. Then G is some common divisor of Q and Q', and we set $\beta = (\delta + p)/q$.

We obtain the four equations

$$\begin{aligned}
A - \beta A' &= G\,, \\
B - \beta B' &= -G\alpha' = -G(\delta + P')/Q'\,, \\
C - \beta C' &= -G\alpha \ = -G(\delta + P)/Q\,, \\
D - \beta D' &= G\alpha\alpha' = G(\delta + P)(\delta + P')/QQ'
\end{aligned}$$

and their conjugates. With little more effort than recalling that

$$\delta^2 - t\delta + n = 0$$

we find

$$\begin{aligned}
A' &= 0\,, & A &= G\,, \\
B' &= Gq/Q'\,, & B &= G(p - P')/Q'\,, \\
C' &= Gq/Q\,, & C &= G(p - P)/Q\,, \\
D' &= -G\frac{q(P + P' + t)}{QQ'}\,, & D &= G\frac{PP' - n - p(P + P' + t)}{QQ'}\,.
\end{aligned}$$

It remains to see whether we can solve this system for G a common factor of Q and Q' and for p and q; with integers $A, \dots, A', \dots$, not all sharing a common factor.

It seems natural to try simple things first and to ensure that $A, \dots, D$, $A', \dots, D'$ — except perhaps D and D' — be integers by trying

$$q = QQ'/G^2\,, \qquad p \equiv P \pmod{Q/G}\,, \quad \text{and} \quad p \equiv P' \pmod{Q'/G}.$$

Here we should note that the conditions on p do not quite determine p modulo q. Indeed, if the greatest common divisor (Q, Q') is F and $F = GH$, then we have only fixed p modulo q/H.

Turning next to D', we see that its wish to be an integer demands that $G \mid (P + P' + t)$. That suggests our fixing G as the greatest common divisor $G = \gcd(Q, Q', P + P' + t)$. Finally, we consider D, wisely observing that it can be reexpressed as $D = G((P - p)(P' - p) - (n + tp + p^2))/QQ'$. We know *a priori* — by virtue of our notation — that

$$n + tp + p^2 \equiv 0 \pmod{QQ'/G^2}$$

since $q = QQ'/G^2$. But we have retained the freedom to fix p so that not just $(P - p)(P' - p) \equiv 0 \pmod{QQ'/G^2H}$ but that moreover

$$\frac{(P - p)(P' - p)}{QQ'/G^2H} \equiv 0 \pmod{H}\,.$$

Thus we can select p so that d is an integer; we do, and we are done. Just. In summary, we select

$$\begin{aligned}
&G = \gcd(Q, Q', P + P' + t)\,, \\
&q = QQ'/G^2\,, \\
&p \equiv P \pmod{Q/G}\,, \quad \text{and} \quad p \equiv P' \pmod{Q'/G}
\end{aligned}$$

with $(P-p)(P'-p) \equiv 0 \pmod{QQ'/G^2}$.

These selections for G, p, q yield the integrality of all of $A, \ldots, D, A'$, $\ldots, D'$. So we have a solution to our system of equations and have found a rule for compounding the given forms, that is, provided we can cope with arbitrary K and K'. But we can do that by repeating the preceding argument with $\alpha = (\delta + K'P)/K'Q$ and $\alpha' = (\delta + KP')KQ'$ and with $\beta = (\delta + KK'p)/KK'q$. We will have done nothing dramatic other than to replace $KK'\delta$ by δ. We must therefore select

$$G = \gcd(K'Q, KQ', K'P + KP' + KK't),$$
$$q = KK'QQ'/G^2,$$
$$Kp \equiv P \pmod{Q/G}, \quad \text{and} \quad K'p \equiv P' \pmod{Q'/G}$$

with $(P-Kp)(P'-K'p) \equiv 0 \pmod{QQ'/G^2}$.

There are some final bits of tidying we should undertake, such as noting that any common factor of $A, \ldots, A', \ldots$ arises from the coefficients of one or both the forms

$$Q(x-\alpha y)(x-\overline{\alpha}y) \quad \text{and} \quad Q'(x'-\alpha'y')(x'-\overline{\alpha}'y')$$

having a common factor. We settle that problem by noting that we could always avoid the forms not being *primitive* by a sensible choice of the parameters defining the canonical form of α and α'.

XVI.6 We should also agree that it was so easy to move back to the general case of K and K' that it is almost spurious to clutter up the formulas with K s at all. Therefore below, we don't. Let's also choose to relax for a moment and, seemingly apropos of nothing, compute the six determinants

$$\Delta_{A,B} = \begin{vmatrix} A & B \\ A' & B' \end{vmatrix}, \quad \Delta_{B,C} = \begin{vmatrix} B & C \\ B' & C' \end{vmatrix}, \quad \ldots .$$

We might set out our working as an array

$$\begin{matrix} G & G(p-P')/Q' & G(p-P)/Q & G\big((p-P)(p-P') - (n+pt+p^2)\big)/QQ' \\ 0 & Gq/Q' & Gq/Q & -Gq(P+P'+t)/QQ' \end{matrix}$$

and proceed to compute the determinants of its six maximal minors. We obtain simple expressions

$$\Delta_{A,B} = Q, \quad \Delta_{D,A} + \Delta_{B,C} = 2P + t, \quad \Delta_{C,D} = (n + tP + P^2)/Q$$
$$\Delta_{A,C} = Q', \quad \Delta_{D,A} - \Delta_{B,C} = 2P' + t, \quad \Delta_{B,D} = (n + tP' + P'^2)/Q'.$$

What fun! These are the six coefficients of the forms we composed. That cannot be an accident nor is it just a property of the particular form of compounding we are discussing.

My description of composition may have been distracting in that it leaves the illusion that we found the unique way to compound a pair of forms. That is not so. To understand all we return to the "cute remark" above whereby we were led to observe that $\Phi(X, Y)$, viewed as a form in x and y, both has discriminant

$$\begin{vmatrix} Ax' + By' & Cx' + Dy' \\ A'x' + B'y' & C'x' + D'y' \end{vmatrix}^2 (v^2 - 4wu)$$

and discriminant $(\varphi'(x', y'))^2(V^2 - 4WU)$. Thus, indeed, whenever we compound, we do obtain

$$\varphi' = \Delta_{A,C}\, x'^2 + (\Delta_{D,A} - \Delta_{B,C})x'y' + \Delta_{B,D}\, y'^2$$

and similarly, of course,

$$\varphi = \Delta_{A,B}\, x^2 + (\Delta_{D,A} + \Delta_{B,C})xy + \Delta_{C,D}\, y^2.$$

This is not quite plain; the argument only shows this up to multiplication by constants. However, if the coefficients of both φ and φ' have no common factor — in the old languages, if they are primitive or if their content is 1 — then all is well. Whatever, we now know that compounding has the six determinants as its invariants.

That reveals all. An array

$$\begin{matrix} A & B & C & D \\ A' & B' & C' & D' \end{matrix}$$

defines a projective 1-manifold in 3-space by way of its *Grassmannian*, namely exactly the six projective coordinates — the six determinants — I have been mentioning. A congenial reminder of such things can be found in Klein's remarks, *Elementary mathematics from an advanced standpoint: Geometry*, reprint (New York: Dover, 1939). I knew about this viewpoint from that source, as well as from more technical sources such as W. V. D. Hodge and D. Pedoe, *Methods of Algebraic Geometry* (Cambridge: Cambridge University Press, 1953). The array thus defines the manifold uniquely up to invertible 2×2 transformations not disturbing those six determinants, that is, up to transformations

$$X' = aX + bY, \qquad Y' = cX + dY \quad \text{with} \quad ad - bc = \pm 1$$

of the compound Φ.

XVI.7 Such transformations are exactly the *equivalence transformations* of Φ. So compounding is well defined only up to equivalence. Moreover, my remarks make it easy to see that if we were to replace ϕ or ϕ' by an equivalent form, then the equivalent forms would compound to yield a form equivalent to the original composite Φ.

XVI.8 So we have learned that compounding is well defined if viewed as an operation on equivalence classes of forms of the same discriminant. Since it's easy to see that the principal form

$$x^2 - txy + ny^2$$

belongs to the identity class, and forms

$$Ux^2 \pm Vxy + Wy^2$$

are representatives of inverse classes, we see that the equivalence classes form a group with respect to compounding. This is essentially the class group whose order $h(\Delta)$ we have been speaking about. Note incidentally that $\Delta = t - 4n^2$.

XVI.9 One should convince oneself that when there are inequivalent forms of some discriminant Δ, that is if the class number is greater than 1, then the ring of integers of $\mathbb{Q}(\sqrt{\Delta})$ can no longer have unique factorization. A hint is that a prime $p \in \mathbb{Z}$ splits in $\mathbb{Q}(\sqrt{\Delta})$ if and only if it is represented by forms of the principal class.

XVI.10 It's probably not obvious that the collection of equivalence classes of forms of discriminant Δ, with compounding as its operation, is essentially the same as the ideal class group of the field $\mathbb{Q}(\sqrt{\Delta})$. The idea is to think of a form $Qx^2 - (2P + t)xy + Q'y^2$ as simply disguising the element $(\delta + P)/Q$ and thence the $\mathbb{Z}$-module $\langle Q, \delta + P\rangle$ — all $\mathbb{Z}$-linear combinations of Q and $\delta + P$. It only remains to verify, cleverly using that $Q \mid (\delta + P)(\overline{\delta} + P)$, that this $\mathbb{Z}$-module is stable under multiplication of algebraic integers in $\mathbb{Q}(\sqrt{\Delta})$, and thus is an ideal of the ring of integers of that field. Equivalent forms induce equivalent ideals; composition details the manner in which the $\mathbb{Z}$-modules multiply; and there we are. Mind you, there remains plenty of opportunity to become confused by such problems as just which of its "zeros" a form corresponds to; or that the ideals $\langle \pm Q, \delta + P\rangle$ are the same; or whether we should mean just proper equivalence by "equivalence" or .... An attempt to explain these things is made in my remarks with Richard Mollin, 'A note on symmetry and ambiguity' [*Bull. Austral. Math. Soc.*, **51** (1995), pp. 215–233].

XVI.11 A simplification and readable summary of Goldfeld's work is given by Joseph Oesterlé in 'Nombres de classes de corps quadratiques imaginaires' [Séminaire Bourbaki 36e année, 1983–84, nº 631, in *Astérisque* **121–122** (1985), pp. 309–323].

XVI.12 When $\Delta > 0$, so for indefinite forms

$$Rx^2 - (2S + t)xy - R'y^2 = R(x - \alpha y)(x - \overline{\alpha} y),$$

the idea needed to prove the finiteness of the class number is to study the continued fraction expansion of $\alpha = (\delta + S)/R$. After finitely many steps of the continued fraction algorithm (see Lecture II) that yields a complete quotient $\beta = (\delta + P)/Q$ satisfying $\beta > 1$ and $-1 < \overline{\beta} < 0$. It is immediate that the *reduced* form

$$Q(x - \beta y)(x - \overline{\beta} y) = Qx^2 - (2P + t)xy - Q'y^2$$

is equivalent to the original form and easy to check that $0 \le P < \delta$, $0 < Q < \delta - \overline{\delta}$; so there are only finitely many reduced forms of given positive discriminant. Hence again the class number is finite. However, in the indefinite case it is rather harder to compute the class number, because in general each class contains many different reduced forms; in fact, exactly as many forms as the length of the period of the continued fraction expansion of δ. In this case it seems that there is no lower bound on the class number in terms of the discriminant. But it is certainly not known how to produce infinitely many discriminants $\Delta > 0$ with class number $h(\Delta) = 1$.

XVI.13 The problem is caused by the existence of a nontrivial unit in the ring of integers of $\mathbb{Q}(\sqrt{\Delta})$ when $\Delta > 0$. The unit group has rank 1 and the size of the *regulator* of the field, the logarithm of the absolute value of a generator of the group, of a *fundamental* unit, is closely related to the length of the period. More precisely, if $x + \delta y > 0$ is the fundamental unit, then the continued fraction expansion $x/y = [a_0, a_1, \dots, a_{r-1}]$ entails that we have $\delta = [a_0, \overline{a_1, \dots, a_{r-1}, 2a_0 - t}]$. It is well known that fundamental units can be startlingly large. For example, Fermat knew that the smallest nontrivial solution of $x^2 - 109y^2 = 1$ yields the unit

$$x + \sqrt{109}y = 158070671986249 + \sqrt{109} \cdot 15140424455100 .$$

Mind you, the fundamental unit of $\mathbb{Q}(\sqrt{109})$ is rather smaller, it being a sixth root of $x + \sqrt{109}y$.

XVI.14 I should have remarked earlier in respect of the Birch–Swinnerton-Dyer Conjectures that it is at any rate an easy matter to show that for a modular elliptic curve E, the quantity $L(E, 1)$ is a rational multiple of the real period of the lattice belonging to E. That's clear by integrating from a cusp to a cusp on the appropriate $X_0(N)$. So the point is to describe that rational multiple and in the first instance to confirm that it is zero when E has infinitely many rational points.

XVI.15 Just what is that mysterious Tate–Shafarevitch group? Indeed. A reasonably accessible explanation is given by J. W. S. Cassels in his *Lectures on Elliptic Curves* (Cambridge: Cambridge University Press, 1991). After mentioning the Weil–Châtelet group WC, Cassels darkly footnotes that his "most lasting contribution to the subject is the notation Ш." He adds "The original notation was TS which, Tate tells me, was intended to continue the lavatorial allusion to WC." He concludes by explaining that the contribution of Ш is "the part that is difficult to eliminate."

XVI.16 Barry Mazur's discussion in ['On the passage from local to global in number theory', *Bull. Amer. Math. Soc.*, **29** (1993), pp. 14–50] is very instructive. Let me risk a brief summary. Mazur explains that given, say, a smooth curve V defined over the field of rationals $\mathbb{Q}$, one should ask about the extent to which the "local" curves $V/\mathbb{Q}_p$ — the reductions of V at all the primes p (including the "infinite" prime) — determine the original "global" curve $V/\mathbb{Q}$ (at any rate, up to isomorphism over $\mathbb{Q}$). Mazur suggests we think of a smooth curve $V'/\mathbb{Q}$ as a *companion* to V if $V'/\mathbb{Q}_p$ is isomorphic to $V/\mathbb{Q}_p$ as a curve, for all the p. Denote the set of all such curves V' as $S(V)$. The cardinality $\#S(V)$ of this set is roughly analogous to the class number in that it measures the extent to which the local data fails to determine the global data.

I suggested in Lecture VIII, at least in passing, that the curve

$$V : y^2 = x^4 - 17$$

has $\#S(V) > 1$. Another classical example is Selmer's curve

$$C : 3x^3 + 4y^3 + 5z^3 = 0 .$$

It has, counting itself, exactly five companions; to wit

$$\begin{aligned} 3x^3 + 4y^3 + 5z^3 &= 0 \\ 12x^3 + y^3 + 5z^3 &= 0 \\ 15x^3 + 4y^3 + z^3 &= 0 \\ 3x^3 + 20y^3 + z^3 &= 0 \\ 60x^3 + y^3 + z^3 &= 0 . \end{aligned}$$

This is a highly nontrivial fact. All five curves have genus 1 but only the last has a rational point, so only it is an elliptic curve. Mazur emphasizes that *a priori* there is no guarantee that the companion curves are isomorphic to smooth plane curves over $\mathbb{Q}$, nor is there any *a priori* bound on their degree. Before the work of Rubin and Kolyvagin one could not even begin to establish that $S(C)$ is finite.

All five equations have nontrivial rational solutions over all $\mathbb{Q}_p$ but only the last has a rational solution, namely $(0, 1, -1)$, over $\mathbb{Q}$. Taking this point as the point O, the curve defined by $60x^3 + y^3 + z^3 = 0$ is an elliptic curve E over $\mathbb{Q}$ isomorphic to the Jacobian of each of the five curves on the list.

In an analogy with the class group of quadratic number fields as just discussed, the Tate–Shafarevitch group describes the extent to which the local-global principle fails on a curve C. Just as restricting to proper equivalence may make the class group larger, here too one attaches some extra structure to these curves. In this example each of the first four curves yields two elements of Ш — its order is 9, a square as predicted.

XVI.17 Given a genus 1 curve C over $\mathbb{Q}$, one defines a new curve $J = \mathrm{Jac}(C)$, its Jacobian. On the set $C \times C$ of pairs of points, we define an equivalence relation whereby $(P_1, Q_1) \sim (P_2, Q_2)$ if and only if P_1, Q_2 are the poles and P_2, Q_1 the zeros of a function on C. Then (up to birational equivalence) J is the curve whose points are those equivalence classes of points. J is defined over $\mathbb{Q}$ and certainly contains a rational point, namely the point corresponding to the class (P, P) on $C \times C$. One says that C is a *principal homogeneous space* for J. Clearly, C and $\mathrm{Jac}(C)$ are birationally equivalent if and only if C has a rational point. Because J has a rational point it has a canonical group structure. For a fixed J the principal homogeneous spaces modulo birational equivalence form the Weil–Châtelet group, a commutative torsion group of infinite order. Those C which contain points defined over each $\mathbb{Q}_p$ and $\mathbb{R}$ provide a certain number of equivalence classes comprising the Tate–Shafarevitch group. It was always theoretically possible to compute its finite subgroups consisting of its points of any given order. The work of Rubin and Kolyvagin has allowed more in certain limited cases.

XVI.18 More generally, one can define the Jacobian variety belonging to a curve V of any genus. It has the "universal" property that a map from V to an elliptic curve E must factor through $\mathrm{Jac}(V)$. This is the sense in which the Modularity Conjecture asserts that every elliptic curve of conductor N appears as a factor of the Jacobian of the modular curve $X_0(N)$. Exactly in the cases when $X_0(N)$ has genus 1, namely the cases $N = 11, 14, 15, 17, 19, 20, 21, 24, 27, 32, 36$, and 49, the Jacobian may coincide with $X_0(N)$ (apparently in the case $N = 17$, $X_0(17)$ is a two-covering of $\mathrm{Jac}(X_0(17))$).

XVI.19 John Coates remarks in 'The work of Gross and Zagier on Heegner points and the derivatives of L-series' [*Séminaire Bourbaki*, 37e année, 1984–85, n° 635 in *Asterisque* **133–134** (1986), pp. 57–72], that it had long been known that one can construct solutions of Pell's equation either by using the values of circular functions or using the values of Dedekind's η-function. But the great credit for the first successful attempt using the values of elliptic modular functions in constructing rational points on elliptic curves is due to Kurt Heegner ['Diophantische Approximationen und Modulfunktionen', *Math. Z.* **56** (1952), pp. 227–253], who in the same paper applied similar ideas to give the first effective determination of all imaginary quadratic fields with class number 1. After a period of neglect in obscurity, Heegner's ideas were taken up and extended by Birch.

The importance of that work is that it establishes, for the first time, the existence of rational points of infinite order on certain elliptic curves over $\mathbb{Q}$, without actually writing down the coordinates of those points and naïvely verifying that they satisfy the equation of the curve.

One calls the points so obtained *Heegner points* on the elliptic curve. On studying the Heegner–Birch construction, Birch and Stephens were led to the conjectures that the Heegner points on an elliptic curve are of infinite order if and only if the group of rational points on the curve has rank 1; and if the Hasse–Weil L-series of the elliptic curve vanishes at $s = 1$ there is a closed formula for its derivative as a product of a standard nonzero period term and the canonical height of its Heegner points.

XVI.20 Heegner's proof for the class number 1 problem was believed to have a fatal gap and was not accepted as a proof at the time of its publication. A clear version of the argument is given by Harold Stark in Antwerp I [H. M. Stark, 'Class-numbers of complex quadratic fields', in *Modular Functions of One Variable* I, Lecture Notes in Mathematics **320** (New York: Springer-Verlag, 1973)]; for other details, see H. M. Stark, 'On the "gap" in a theorem of Heegner' [*J. Number Theory* **1** (1969), pp. 16–27].

XVI.21 Fortunately, this sad story is not well known, for otherwise it would fuel the persecution complexes of amateurs whose "proofs" continue to be denied. In brief, Heegner, a schoolteacher (which then meant rather more in terms of mathematical background in Germany than it now does in, say North America or Australia), published new observations on modular functions which apparently, but rather unexpectedly, led to a proof of the class number 1 problem. However, the arguments were sufficiently obscurely written to leave considerable doubt about their completeness, even in essence. Heegner's real contribution is by now well recognized. The question is, was it a disgraceful scandal that his contribution was not recognized in his lifetime?

I think not. An extreme response runs as follows: My family is owned by a large golden retriever named "Talleyrand"†. For all I know, when Tal smiles at me and barks peremptorily, he is not demanding a walk but is describing a simple proof of Fermat's Last Theorem. Whatever, handsome though he is, it seems unlikely that he might have a proof, and in any case his argument is unclear to me. I think it correct not to spend too much time trying to comprehend it. In that spirit, an amateur, or for that matter a recognized mathematician, had best have clear arguments written in the language of the majority — the language expected by other mathematicians — if her surprising arguments are to get a proper hearing. That's not unfair; it's our playing the odds. If in consequence great contributions are neglected, that will be a misfortune, not a scandal.

XVI.22 When the conductor N is squarefree, thus in the semistable case now settled by Wiles, the sign of the functional equation can be determined in purely geometric terms. It depends only on the number of a_p, for primes $p \mid N$ of bad reduction, that are positive. But for those p we have $a_p = 1$ if the tangent directions at the double point are defined over $\mathbb{F}_p$ and $a_p = -1$ otherwise. Specifically, the rank of the Mordell-Weil group is odd if and only if the number of p dividing N for which the tangent directions are defined over $\mathbb{F}_p$ is even.

XVI.23 Zagier explains in his research-expository survey that while computation readily proves that a certain quantity is not zero, computation alone cannot possibly show that it does indeed vanish. Thus to show that $P^{(-139)} = 0$ requires some other argument: The class number of

† The name is inaccurate. Technically, Tal is the 1983 successor of a short-lived predecessor, Talleyrand, and is formally named "Talley II", after the America's Cup winning yacht, Australia II.

$\mathbb{Q}(\sqrt{-139})$ is 3 and for the three points z_j (with $N = 37$ and $\beta = 3$) one may choose

$$\frac{-2 + \frac{1}{2}(1 + i\sqrt{139})}{37}, \quad \frac{35 + \frac{1}{2}(1 + i\sqrt{139})}{5 \cdot 37}, \quad \text{and} \quad \frac{-76 + \frac{1}{2}(1 + i\sqrt{139})}{5 \cdot 37}.$$

These satisfy $37z = (az + b)/(cz + d)$ with

$$\begin{pmatrix} -3 & -1 \\ 1 & 0 \end{pmatrix}, \begin{pmatrix} -77 & -31 \\ 5 & 2 \end{pmatrix}, \begin{pmatrix} 34 & -7 \\ 5 & -1 \end{pmatrix} \in \mathrm{SL}(2, \mathbb{Z}),$$

respectively. In each case $(cz + d)^{-1} = \frac{1}{2}(3 + i\sqrt{139})$. Because of the transformation formula

$$\Delta\left(\frac{az + b}{cz + d}\right) = (cz + d)^{12}\Delta(z)$$

satisfied by the discriminant

$$\Delta(z) = q\prod_{n=1}^{\infty}(1 - q^n)^{24}$$

— recall that as usual q denotes $e^{2\pi iz}$ — it follows that the function

$$\sqrt[12]{\frac{\Delta(z)}{\Delta(37z)}} - (1 + \tfrac{1}{2}(1 + i\sqrt{139}))$$

$$= q^{-3}\prod_{n=1}^{\infty}\left(\frac{1 \quad q^n}{1 - q^{37n}}\right) - (1 + \tfrac{1}{2}(1 + i\sqrt{139}))$$

vanishes at z_1, z_2, and z_3. On the other hand, $g(z)$ is $\Gamma_0(37)$-invariant, it has a triple pole at $z = \infty$, and it has no other poles since $\Delta \neq 0$ in $\mathcal{H}$. So these are the only three zeros. It follows that indeed

$$\phi(z_1) + \phi(z_2) + \phi(z_3) = 0 \in E(\mathbb{C})$$

for any map ϕ, with $\phi(\infty) = 0$, from $X_0(37)$ to an elliptic curve E.

XVI.24 Early remarks on the genesis of the Birch–Swinnerton-Dyer Conjectures make quite interesting reading; a good, accessible, instructive, and comprehensible example is B. J. Birch, 'Conjectures concerning elliptic curves', in *Proceedings of the Symposium on Pure Mathematics* VIII (Providence, R.I.: American Mathematical Society, 1965), pp. 106–112. Nowadays these conjectures are just some special cases of a vast web of conjectures, dealing with more general objects than elliptic curves. A useful description of such things is provided by Wilfred W. J. Hulsbergen, *Conjectures in Arithmetic Algebraic Geometry*, 2nd rev. ed. (Wiesbaden, Germany: Vieweg, 1994).

Notes on Fermat's Last Theorem

LECTURE XVII

When Fermat Vapours clog our loaded Brows,
With furrow'd frowns, when stupid downcast Eyes
Th'external Symptoms of some Gap within
Our Proof express, or when in sullen Dumps
With Head Incumbent on Expanded Palm,
Moping we sit, our Gauloise snuffed, deform'd,
Sing then, Oh Wiles, and Taylor, Wiles!
Oh trio: put Fermata to our Toils.

Barry Mazur*

The first step in the proof of Fermat's Last Theorem is Frey's observation† that if u, v, w are integers satisfying $u^p + v^p + w^p = 0$, then the curve

$$F : y^2 = x(x - u^p)(x + v^p),$$

although elliptic, can be shown to have properties likely to contradict the Modularity Conjecture if the odd prime p is at least 5. Here one arranges, as one may, that $u \equiv -1 \pmod 4$ and that $2 \mid v$.

The Modularity Conjecture asserts that a rational elliptic curve

$$E : f(x, y) = 0$$

is parametrized $f(g_x, g_y) = 0$ by certain modular functions g_x, g_y. If E has conductor N, then these functions are modular with respect to the group $\Gamma_0(N)$ of 2×2 unimodular matrices $\begin{pmatrix} a & b \\ Nc & d \end{pmatrix}$. A natural combination of the two modular functions derived from a holomorphic differential on the Riemann surface $E(\mathbb{C})$ then yields a modular form $f_E(z)$ of weight 2 with Fourier expansion of the shape

$$f_E(z) = \sum_{n=1}^{\infty} a_n q^n, \quad \text{for} \quad q = e^{2\pi i z}$$

with the a_n integers and $a_1 = 1$. In fact, work of Shimura entails that for primes p the a_p yield the number of points $\#E(\mathbb{F}_p)$ on the reduction of

*The generally acknowledged winning entry to the "Fermat's Last Theorem Poetry Challenge" at the Boston University *Fermat's Theorem* meeting, August 1995.

†Notes to Lecture IX; but also several other remarks.

the curve at p by the formula $a_p = p + 1 - \#E(\mathbb{F}_p)$. Arguments deriving from work of Hecke imply that $f_E(z)$ must be an eigenform for all the operators of a certain Hecke algebra, whence one has the Euler product

$$L(E,s) = \sum a_n n^{-s} = \prod_{p|N}(1 - a_p p^{-s})^{-1} \prod_{p\nmid N}(1 - a_p p^{-s} + p^{1-2s})^{-1}$$

yielding the a_n in general. In summary, $f_E(z)$ must be a cusp eigenform of level N; specifically, following observations of Atkin and Lehner, they are *newforms*: these Hecke eigenforms do not also belong to some lower level properly dividing N.

Let's return to the case of Frey curves. The discriminant of the Frey curve is $\Delta_F = ((u^p v^p w^p)/16)^2$; its conductor N_F is given as the product $N_F = \prod_{p|uvw} p$. In particular, the curve is semistable. Its j-invariant is

$$2^8(u^p v^p + v^p w^p + w^p u^p)^3/(u^p v^p w^p)^2.$$

Suppose l is an odd prime dividing N_F. Then l can't divide the numerator of j and so the number of times that l divides j_F, the l-adic order of j_F, is some (negative) integer multiple of p.

These are rather special circumstances in which a conjecture of Serre on "level reduction" for modular galois representations would fill the gap in the argument sketched by Frey. In 1986 Ken Ribet proved this conjecture.

Specifically, suppose that the form $f_F = \sum a_n q^n$ belonging to a Frey curve F is indeed a cusp form of weight 2 and level $N = N_F$. Then level reduction shows that for each l dividing N there is a cusp form f' of weight 2 and level N/l with $f' \equiv f_F \pmod{p}$, and with f' belonging to the same galois representation $\rho_{F,p}$ as did f_F, except it has smaller level N/l.

Two forms are congruent modulo p if their corresponding coefficients a_n, a'_n each satisfy $a_n \equiv a'_n \pmod{p}$; happily, to confirm congruences requires checking just a bounded number of n. Clearly, the reduction process may be repeated, sequentially eliminating each different odd prime divisor of N, until eventually we obtain a cusp form $\tilde{f}$ of weight 2 and level 2; note that since v is even it follows that 2 does divide N. But there are no cusp forms of weight 2 and level 2. Thus the Modularity Conjecture cannot hold for the Frey curve F.

It's now clear that the trick in proving Fermat's Last Theorem is to show that all semistable elliptic curves *are* modular, thereby denying the existence of Frey curves. So somehow, given an elliptic curve E, we need to link it to a modular form. Sadly, that is done in a very indirect way.

Wiles', and everyone else's, approach to the study of elliptic curves is via their associated galois representations. Let me hint at what that might mean before mentioning some of the immediately comprehensible aspects of the proof. Briefly, for different integers m, one studies the collection $E[m]$ of the m-division points on a given elliptic curve E, that is, the points Q on $E(\mathbb{C})$ satisfying $mQ = O$. Generically, there are m^2 of them, and because of the algebraic equations defining their coordinates, they of course lie on $E(\overline{\mathbb{Q}})$, where $\overline{\mathbb{Q}}$ is the field of all algebraic numbers. Moreover, $E[m]$ is a subgroup of $E(\mathbb{C})$ and the remark just made confirms it is already a subgroup of $E(\overline{\mathbb{Q}})$. Indeed, the coordinates of the m^2 points of $E[m]$ generate a finite extension of $\overline{\mathbb{Q}}$, a number field referred to as K_m below. It is now natural to ask how the elements of $E[m]$ permute under the automorphisms σ of $\overline{\mathbb{Q}}$, that is, under the action of the galois group $G = \mathrm{Gal}(\overline{\mathbb{Q}}/\mathbb{Q})$. Since $E[m]$ is itself a subgroup of $E(\overline{\mathbb{Q}})$, the group of algebraic points of $E(\mathbb{C})$, the action of a σ on $E[m]$ is just an element of $\mathrm{Aut}(E[m])$, the group of automorphisms of $E[m]$. Since $E[m]$ is isomorphic to the group $(\mathbb{Z}/m\mathbb{Z})^2$, it follows that $\mathrm{Aut}(E[m])$ is isomorphic to the group $\mathrm{GL}(2, \mathbb{Z}/m\mathbb{Z})$ of 2×2 matrices with integer entries modulo m.

Since the elements of $E[m]$ are permuted by the galois group, it follows that the field K_m generated by their coordinates over $\mathbb{Q}$ is a finite galois extension of $\mathbb{Q}$. Finally then, we may view the action of $\mathrm{Gal}(\overline{\mathbb{Q}}/\mathbb{Q})$ on $E[m]$ as a continuous homomorphism

$$\rho_{E,m} : \mathrm{Gal}(\overline{\mathbb{Q}}/\mathbb{Q}) \to \mathrm{Aut}(E[m])$$

and think of $\rho_{E,m}$ as embedding the subgroup G_m of the galois group $\mathrm{Gal}(K_m/\mathbb{Q})$ as a subgroup of the matrix group $\mathrm{GL}(2, \mathbb{Z}/m\mathbb{Z})$. This is the *galois representation.* Serre has shown that if E does not have complex multiplication, then $G_p = \mathrm{GL}(2, \mathbb{F}_p)$ for all but finitely many primes p.

The point of all this is that the representations $\rho_{E,m}$ contain critical arithmetic information about the curve E. The discriminant of the field K_m happens to be divisible only by primes l dividing either m or the conductor N of the elliptic curve E. So for $l \nmid mN$ one can introduce a Frobenius element σ_l in G_m — corresponding to the automorphism given by replacing elements by their l th power — and view σ_l as a matrix whose trace and determinant are well-defined elements of $\mathbb{Z}/m\mathbb{Z}$. The determinant is just l modulo m; but, critically, we have the congruence

$$\mathrm{Trace}(\sigma_l) \equiv a_l \pmod{m}$$

where a_l is the number $l + 1 - \#E(\mathbb{F}_l)$.

This is a theorem of Deligne and Serre; many cases had already been established by Shimura. So the representation $\rho_{E,m}$ determines numbers $a_l \pmod{m}$ for almost all l, that is, all l not dividing mN. It follows that properties such as the conjectured modularity of the form $\sum a_n q^n$ arising from the L-series $\sum a_n/n^s$ translates to a property — naturally also called "modularity"— of the representations $\rho_{E,m}$.

One determines modularity by confirming that f is simultaneously an eigenform for certain endomorphisms of Fourier series, the Hecke operators* T_n. These T_n commute with one another and are interrelated by identities which express a given T_n in terms of the Hecke operators

*See the Notes following Lecture XII.

indexed by the factors of n. Specifically, if $l \nmid N$, then $f|T_l$ (this being the usual notation for the form $T_l f$) is given by $\sum a_{nl}q^n + l\sum a_n q^{ln}$; if $l|N$, it is just $\sum a_{nl}q^n$. Generally,

$$f|T_n = \sum_{k=1}^{\infty}\Big(\sum_{\substack{(d,N)=1\\ d|(k,n)}} d a_{kn/d^2}\Big)q^k .$$

But there is and was no program in sight that might do that, or that might somehow link modular forms to the representations $\rho_{E,l}$ for the different primes l. However, there is the alternative of taking just the one prime l and studying the family of sets $E[l^k]$, $k = 1, 2, \ldots$. The resulting sequence of representations

$$\rho_{E,l^k} : \mathrm{Gal}(\overline{\mathbb{Q}}/\mathbb{Q}) \to \mathrm{GL}(2, \mathbb{Z}/l^k\mathbb{Z})$$

can then be repackaged as a single representation

$$\rho_{E,l^\infty} : \mathrm{Gal}(\overline{\mathbb{Q}}/\mathbb{Q}) \to \mathrm{GL}(2, \mathbb{Z}_l)\,,$$

where $\mathbb{Z}_l$ is the ring of l-adic integers. The happy thing here is that only a single prime l is involved. In that case there is a result of Langlands-Tunnell yielding that the representations $\rho_{E,l}$ are modular if l is at most 3.

So Wiles now looks for an argument allowing him to lift the modularity of $\rho_{E,l}$ for odd primes l, of course particularly for $l = 3$, but also $l = 5$ as it turns out, to that of ρ_{E,l^∞}. The idea is to attempt to disregard E and to work not so much with ρ_{E,l^∞} as such as with l-adic representations with properties generalizing those of ρ_{E,l^∞}. It was here that the "gap" occurred.

In any case, some extra ingenuity is required even presuming success in this modularity lifting argument. The argument certainly requires that the representation $\rho_{E,l}$ be irreducible — meaning that the representation not be the direct sum of representations of smaller dimension (that is, of dimension 1 in the present case). The irreducibility of $\rho_{E,3}$ can easily be checked in any particular example but can't be guaranteed in general, so Wiles is forced, after all, to work with a second prime. But now he does not have the Langlands-Tunnell theorem!

Wiles' ingenious idea around this is as follows. First, one can list the elliptic curves E for which both $\rho_{E,3}$ and $\rho_{E,5}$ are reducible. They are four curves of conductor 50. Checking Antwerp IV, or Cremona, confirms that all are modular. So without loss of generality Wiles may suppose that $\rho_{E,5}$ is irreducible. However, Langlands-Tunnell is not available for $l = 5$.

What Wiles does is to concoct a new elliptic curve E' so that both $\rho_{E',5}$ is isomorphic to $\rho_{E,5}$, *and* $\rho_{E',3}$ is irreducible. The concoction is ingenious and the invocation of the Hilbert Irreducibility Theorem excited onlookers at the time. Now the irreducibility of $\rho_{E',3}$ entails the modularity of E' via Langlands-Tunnell and then the isomorphism at $l = 5$ confirms the modularity of the original curve E.

Mind you, even while the gap remained open it was clear that Wiles had achieved a great deal. His partly successful argument dealing with E semistable, l an odd prime and $\rho_{E,l}$ irreducible, provided techniques in any case able to exhibit an infinite class of semistable elliptic curves over $\mathbb{Q}$ that are modular. No such class was known before Wiles' methods. Moreover, in the special case that the representation $\rho_{A,l}$ arises from a complex multiplication elliptic curve A over $\mathbb{Q}$ there was no "gap" in the proof; once again the result can be used to provide an infinite collection of elliptic curves over $\mathbb{Q}$ that are modular.

We now know that Wiles, with some aid from Richard Taylor, has been able to fill in the necessary details. Those details don't fit well into the story I've been telling here and in my previous lectures; at first they seem to be a chapter of commutative algebra rather than about elliptic curves as such. Ken Ribet's talk at the Minneapolis Mathfest, August 1994 on 'Galois representations and modular forms', provides many useful remarks augmenting the description in Karl Rubin and Alice Silverberg's 'A report on Wiles' Cambridge lectures'* [*Bull. Amer. Math. Soc.*, **31** (1994), pp. 15–38]. Naturally I have relied strongly on these reports.

I recall, from back in 1987, Tate mentioning Serre's explanation that the Frey–Ribet argument plus the Modularity Conjecture would lead to settling a somewhat wider class of diophantine equations than just that of Fermat. These include equations of the shape $x^p + y^p = l^k z^p$, where l and p are different primes. The idea is of course to construct the Frey curve belonging to a putative solution, to lower the level of the associated modular form by Ribet's theorem, and then to study the resulting space of modular forms. If there are no forms there is no solution. In particular, Fernando Gouvêa ['A marvelous proof', *Amer. Math. Monthly* **101** (1994), pp. 203–222] remarks that this entails there is no solution for $l = 3, 5, 7, 11, 13, 17, 19, 23, 29, 53$, or 59 provided that p is at least 11 and more generally, provided that $l \neq 2^n \pm 1$, if p is sufficiently large relative to l.

Notes and Remarks

XVII.1 Ribet's proof relates Wiles' work to Fermat's Last Theorem. In principle, Wiles might have proved much of the Modularity Conjecture for its own sake and then have been startled when a remark of Frey and a theorem of Ribet proved the FLT. It didn't happen that way. Indeed, some of Ribet's arguments and not just his conclusion are needed by Wiles for his proof. And, moreover, Wiles has confessed that his motivation really was to prove Fermat's Last Theorem.

*This was originally titled 'Wiles' proof of Fermat's Last Theorem', but then the "gap" intervened.

XVII.2 In the final lecture of the Fermat Meeting at Boston University [August 1995] Wiles remarked that, until Ribet's work, it was naturally considered unprofessional for a mathematician to spend serious time on Fermat's Last Theorem. He then thanked Ken Ribet for "giving me the excuse to spend so many years on my favorite problem in a professional context." Frey too was thanked for hinting at the link; indeed, Wiles concluded by suggesting that Frey might care to make a passing remark about "those zeroes ... ".

XVII.3 It seems that many are rather irritated that the FLT has been settled (and were rather pleased by the "gap"). But science marches on; records get broken. I recommend the ABC-conjecture* as the next "Holy Grail".

XVII.4 In the meantime, briefly detailing the proof does call for a fair amount of arm-waving; but the details do rapidly become less fearsome.

With a little ingenious phrasing,
The proof's detail is no longer dazing.
It's enough just to dream
Of a finite flat scheme
And to say that the proof is amazing.

XVII.5 Whatever, one should feel no sympathy for those mathematicians who think it "not fair" that the FLT has been settled by methods they do not understand. The underlying ideas are perfectly accessible and mostly are part of what should be — I'm sure much of it will be — mainstream mathematics. The details will be simplified and soon they too will become readily comprehensible. It all will be obvious as in Martin Huxley's

A theorem apparently new
Has long been believed to be true
It's implicit in Gauß
And some work of Landau's
That unity doubled is two.

XVII.6 Apropos of the alternative tack Wiles follows in his argument: I felt I first began to understand the notion of local to global arguments when I appreciated that to prove some integral quantity is zero often is done elegantly by showing it to be divisible by arbitrarily many different primes. Alternatively, it is enough to show it to be divisible by arbitrarily high powers of just the one prime.

XVII.7 The "repackaging" spoken of above collects the $E[l^k]$ as the *Tate module* $\mathrm{Ta}_l E$. An onto morphism $\psi : E \to E$ induces an endomorphism ψ^* of $\mathrm{Ta}_l E$, so an element of $\mathrm{Aut}(\mathrm{Ta}_l E) \cong \mathrm{GL}(2, \mathbb{Z}_l)$. Then $\det(\psi^*) = \deg(\psi)$, where, as described in remarks following Lecture XIII, the degree of a map is the number of points (counted according to multiplicity) sent to

*See Lecture XIV.

the identity element O. So $\det(\psi^*)$, which *a priori* is just an element of $\mathbb{Z}_l$, is in fact a rational integer and it is independent of l. In particular, the Frobenius morphism $\phi = \phi_p$ acting on $E(\mathbb{F}_p)$ induces the Frobenius endomorphism $\phi^* = \sigma_p$ acting on the Tate module. One readily finds $\mathrm{Trace}(\sigma_p) = p + 1 - \#E(\mathbb{F}_p)$, much as I briefly described when talking about Hasse's theorem in the Notes to Lecture XIII.

XVII.8 It hasn't until now been necessary to allude to the notion that the modular curve $X_0(N)$ may be viewed as a moduli space. In simpler words: Modular curves are catalogues of elliptic curves plus additional structure. The rough idea is as follows. We know that the compactification of the upper half-plane $\mathcal{H}$ modulo the group $\mathrm{PSL}(2,\mathbb{Z})$, thus $X_0(1)$, corresponds to isomorphism classes of elliptic curves. This is so because different τ in the fundamental domain provide distinct $j(\tau)$ and hence correspond to isomorphism classes of elliptic curves. Each curve has a rational point, namely at ∞. This is the spirit in which it turns out that the *rational* points on $X_0(N)$ correspond to elliptic curves defined over $\mathbb{Q}$ which also possess a rational cyclic subgroup of order N. This interpretation is given by sending a point $\tau \in \mathcal{H}/\Gamma_0(N)$ to the pair $(\mathbb{C}/\langle 1,\tau\rangle, \langle 1/N\rangle)$.

Incidentally, a subset of points is called *rational* if, as a *collection*, the subset can be defined rationally; it's not at all necessary that all the points of the subset be rational. In this sense an algebraic number *together with* all its conjugates provides a rational set because the defining polynomial of the algebraic number is a polynomial over $\mathbb{Z}$.

XVII.9 To find all elliptic curves E for which both $\rho_{E,3}$ and $\rho_{E,5}$ are reducible, one lists the elliptic curves over $\mathbb{Q}$ with a subgroup of order 15 stable under the action of $\mathrm{Gal}(\overline{\mathbb{Q}}/\mathbb{Q})$. But, as just remarked, a modular curve $X_0(N)$ can be viewed in such a way that its rational points other than its cusps correspond to isomorphism classes over $\mathbb{C}$ of pairs (A,C) consisting of an elliptic curve $A/\mathbb{Q}$ and a cyclic subgroup $C \in A(\overline{\mathbb{Q}})$ stable under $\mathrm{Gal}(\overline{\mathbb{Q}}/\mathbb{Q})$ and of order N. Of course, a group of order 15 is cyclic, so for the case needed by Wiles one considers points on the curve $X_0(15)$. It is of genus 1 and is given by the equation $y^2 = x(x+9)(x-16)$. Of the eight rational points on this curve four are cusps. The other four rational points correspond to nonisomorphic elliptic curves. They're of conductor 50, so they are not semistable. Nonetheless, they are modular, coming from $X_0(50)$, as can easily be checked from the tables of modular elliptic curves.

XVII.10 Let me try to detail, following Rubin and Silverberg's description, Wiles' concoction of an elliptic curve E' which is to have $\rho_{E',5}$ isomorphic to $\rho_{E,5}$ *and* $\rho_{E',3}$ irreducible. I set $G = \mathrm{Gal}(\overline{\mathbb{Q}}/\mathbb{Q})$. Wiles appeals to a number of interesting principles. The first is the Hilbert Irreducibility Theorem, which, in one version, says that if a polynomial $P(x, t_1, \ldots, t_n)$ over $\mathbb{Q}$ is irreducible in $\mathbb{Q}[x, t_1, \ldots, t_n]$, then there are plenty of choices of n-tuples $(c_1, \ldots, c_n)$ of rationals so that $P(x, c_1, \ldots, c_n)$ is irreducible in

$\mathbb{Q}[x]$. A second is that congruences modulo powers of different primes, say l and 5, always have a solution — indeed this is no more than the Chinese Remainder Theorem. Finally, and fairly evidently, if two polynomials are p-adically close to one another, then they split in the same way modulo p.

Wiles starts with the classical modular curve $X(5)$. It is just the upper half-plane $\mathcal{H}$ modulo the group $\Gamma(5)$ of unimodular matrices congruent modulo 5 to the identity matrix, plus its cusps. A suitable twist X of this space then has the property that its rational points other than its set of cusps S correspond to isomorphism classes of pairs (E', ϕ) consisting of an elliptic curve $E'/\mathbb{Q}$ and a G-module isomorphism $\phi : E[5] \to E'[5]$. Viewed as a complex manifold $X - S$ is four copies of $\mathcal{H}/\Gamma(5)$, and the component X^0 of X containing the rational point corresponding to $(E, 1)$ is of genus 0, so it has infinitely many rational points.

The idea now is to show that infinitely many of those points correspond to a semistable curve E' with $\rho_{E',3}$ irreducible; that is, E' must not have a G-stable subgroup C of order 3. Hence consider a modular curve $\hat{X}$ with a finite set $\hat{S}$ of cusps so that the rational points on $\hat{X} - \hat{S}$ correspond to isomorphism classes of triples (E', ϕ, C) with (E', ϕ) as above and C a G-stable subgroup of E' of order 3. As a complex manifold $\hat{X} - \hat{S}$ is four copies of $\mathcal{H}/(\Gamma(5) \cap \Gamma_0(3))$, and the map that forgets the subgroup C induces a surjective morphism $\theta : \hat{X} \to X$ defined over $\mathbb{Q}$ and of degree

$$[\Gamma(5) : \Gamma(5) \cap \Gamma_0(3)] = 4.$$

Now let $\hat{X}^0$ be the component of $\hat{X}$ which maps to X^0. Then the function field of X^0 is $\mathbb{Q}(t)$ and the function field of $\hat{X}^0$ is $\mathbb{Q}(t)[x]/f(t,x)$, where we have that $f(t,x) \in \mathbb{Q}(t)[x]$ is irreducible and has degree 4 in x.

By the Hilbert Irreducibility Theorem, irreducibility of $f(t,x)$ entails existence of infinitely many $c \in \mathbb{Q}$ so that $f(c,x)$ is irreducible in $\mathbb{Q}[x]$. Moreover, we can choose a prime $l \neq 5$ such that $f(c,x)$ has no zeros modulo l, and if $c' \in \mathbb{Q}$ is sufficiently close l-adically to c, then $f(c',x)$ has no rational roots. Then c' corresponds to a rational point of X^0, which is not the image of a rational point of $\hat{X}^0$. Thus it corresponds to an elliptic curve E' which has $E'[5] \cong E[5]$ as G-modules, and such that $E'[3]$ has no subgroup of order 3 which is stable under G. Finally, by also choosing c' so that it is sufficiently close 5-adically to the value of t corresponding to E, we also guarantee that E' is semistable at 5.

By the way, it is a sort of sadness that one can avoid the appeal to Hilbert Irreducibility by remarking that the values c to be omitted correspond to certain rational points on a curve of genus 9. Then by Faltings' Theorem there are at most finitely many of them.

XVII.11 Jeremy Teitelbaum* conveniently summarizes the proof:

We take an elliptic curve E,
consider the points killed by 3,
This ρ *must be modular,*
and by facts which are popular,
the proof of Fermat comes for free.

XVII.12 As an example of Wiles' theorem concerning modular elliptic curves arising from a complex multiplication elliptic curve, Rubin and Silverberg cite the family

$$E_t : y^2 = x^3 - x^2 + a_4(t)x + a_6(t)\,,$$

where

$$a_4(t) = -2430t^4 - 1521t^3 - 396t^2 - 56t - 3\,,$$
$$a_6(t) = 40824t^6 + 31104t^5 + 8370t^4 + 540t^3 - 148t^2 - 14t - 1\,.$$

Set $E_0 = E$ noting that the curve $E : y^2 = x^3 - x^2 - 3x - 1$ has complex multiplication by $\mathbb{Q}(\sqrt{-2})$ and has good reduction at 3. For every $t \in \mathbb{Q}$, the representations $\rho_{E_t,3}$ and $\rho_{E,3}$ are isomorphic, and for $t = 3a/b$ or $3a/b + 1$ with $a, b \in \mathbb{Z}$ and $3 \nmid b$, the curve E_t, and any curve isomorphic over $\mathbb{C}$ to E_t, thus with the same j-invariant, is modular.

XVII.13 The papers of Andrew Wiles, 'Modular elliptic curves and Fermat's Last Theorem', and of Richard Taylor and Andrew Wiles, 'Ring-theoretic properties of certain Hecke algebras', are slated to appear in the May 1995 issue of *Annals of Mathematics*. In the meantime the long proof has been "imploding", as study reveals possible simplifications and some generalizations of the arguments. For example, Diamond has now strengthened Wiles' method to entail modularity for all elliptic curves over $\mathbb{Q}$ that do not have additive reduction at 3 and 5. An excellent detailed version of Wiles' proof, 'Fermat's Last Theorem' by Henri Darmon, Fred Diamond and Richard Taylor, together with extensive "introductory" material — assuming almost everything in this book though, and presupposing rather greater mathematical sophistication — is currently [June 1995] circulating in a draft of 113 pages. It is intended to be completed by the end of October, 1995. [August 1995] The preliminary paper appears as 'Fermat's Last Theorem', *Current Developments in Mathematics*, 1995, 1–107.

XVII.14 The disclaimer to those notes is worth repeating. It says *inter alia* that "These notes were prepared in conjunction with a series of lectures given by the authors in the seminar 'Current Developments in Mathematics' held in Boston on 7–8 May 1995. They should be taken in the spirit of an informal transcript of these talks; they are still incomplete,

---

*The person guilty of invoking an outburst of poetry, partly reported in this book, by initiating a Poetry Challenge at the Boston University *Fermat's Theorem* meeting; I found the complete collection of entries at `http://zariski.math.uic.edu/~jeremy`.

and may contain over-simplifications*, inaccuracies, and obscurities of exposition."

XVII.15 This book consists of a very informal transcript of talks that were, or might have been, given at Macquarie University, Sydney. It is quite incomplete; it contains gross over-simplifications, probable inaccuracies, and, sadly, no doubt many obscurities of exposition.

Coda

There is an epilogue. The search is over. Yet mathematics will be nourished by new questions; and new methods will be invented, and they will in turn be applied to other problems ... .†

The papers providing the proof of Fermat's Last Theorem appear as: Andrew Wiles, 'Modular elliptic curves and Fermat's Last Theorem' [*Ann. Math.* (ser. 2) **141** (3), pp. 443–551 (received October 14, 1994)], and Richard Taylor and Andrew Wiles, 'Ring-theoretic properties of certain Hecke algebras' [*ibid.*, pp. 553–572 (received October 7, 1994)]. There is an appendix (received on January 26, 1995) to the latter joint-article dealing with simplifications the authors attribute to Gerd Faltings.

---

*A final spelling check revealed that I had mistyped this as "over-simplications". I nearly retained it as a wonderful instance of a word that described itself. But I was quoting, so I reverted to the original.

†See the Epilogue to Paulo Ribenboim, *13 Lectures on Fermat's Last Theorem* (New York: Springer-Verlag, 1979).

Appendix A

Remarks on Fermat's Last Theorem

I have often thought it might be amusing to write a humorous essay on how to recognize the Dark Ages when you are in them.

In 1637 or so, the French jurist Pierre de Fermat scribbled a remark in the margin of his *Arithmetica* of Diophantus. He wrote that to split a cube into two cubes, or a fourth (biquadratic) power into two fourth powers, or indeed any higher power unto infinity into two like powers, is impossible and that he has a marvelous proof for this. But the margin is too narrow to contain it. [Actually, he wrote: *Cubum autem in duos cubos, ... Hanc marginis exiguitas non caperet*. But it was in another country, and besides, the language is dead.]

By the way. John Coates, Sadleirian Professor of Pure Mathematics at Cambridge — but an Australian, from Possum Brush, NSW — told me a nice story the other day in Hong Kong. He had actually seen, in the library of Emmanuel College, a copy of Bachet's *Diophantus* of 1621. He had been quite startled by the extraordinary width of its margins.

As luck would have it, Fermat's son Samuel was proud of his papa and on dad's death in 1665 proceeded to collect Fermat's mathematical correspondence. That, and a reprint of the *Diophantus* — together with Fermat's marginal notes — was published in 1670. By the end of the eighteenth century all of Fermat's other remarks had been dealt with, one way or the other: either properly proved or shown to be false. Only this one remark, hence the *last* theorem, remained. The story of Fermat's Last Theorem has been told so often it hardly bears retelling. Still, I have begun it and will say more.

These remarks were occasioned by the particularly inane reports that had appeared in the newspapers at the time of Wiles' 1993 announcement of his "proof"; and were written during Christmas 1993 (as a way of my avoiding too much of the festivities).

The present material appeared in the *Australian Mathematical Society Gazette*, **21** (December, 1994), pp. 150–159 with the note that its observations were extracted from my book *Notes on Fermat's Last Theorem* to be published by Wiley-Interscience in 1996.

The social life of a mathematician is not easy. It's an effort to keep on smiling. Apparently competent friends, and strangers, hasten to confess that "I was hopeless at mathematics." Nowadays I try to allay their embarrassment. I say, "Not to worry. I had no alternative to becoming a mathematician. I never learned to read without moving my lips. My secretary writes my letters for me. And frankly, except on Wednesdays when I'm still fresh after my weekend, he kindly reads me my correspondence." [Actually, my secretary is female — my apologies, Kaye — but I thought it sexist to use the automatic "she".]

The psychopathology of the community's reaction to mathematics warrants deeper investigation. Nonetheless, let me offer a few remarks. The stuff that's called "mathematics" in our schools, isn't. It's as far removed from dinkum mathematics, as writing poetry is distant from learning to shape the letters of the alphabet. And, I'm sorry to have to say it, but — maintaining that analogy — an undergraduate degree may mean little more than having learned to write a few grammatically correct sentences. Poetry? If only.

However, it *is* our fault. One can appreciate poetry without being able to write it. Let me try to narrate one of our poems.

Its subject is number theory, the physics of the ideal world of numbers. This study is both easier than, and more subtle than, the physics of the natural world. The objects of my world are quite concrete. But my world is infinite. The physical universe probably is not. We can *prove* that some of the things we say are so. In real physics one can do no more than refine the theories and show by successful prediction that one is on a sensible track. Very loosely speaking, my world is *a priori*. It just is the way it is because it is that way. The real world is *a posteriori*. It kind of seems to be the way it appears to be because it appears, at least from our present viewpoint, to seem that way. Planes fly, telephones ring, bombs explode, average lifespans increase. The scientists don't have it all wrong.

Number theory certainly existed in prehistory when science, mathematics and philosophy were one and the same thing. It seems the human condition to be intrigued by pattern. Every second number is even, it's exactly divisible by 2. An odd number times an odd number is always odd; but their sum is always even. Behold $6 = 1 + 2 + 3$; it's the sum of its own divisors. No wonder that the world was created in this "perfect" number of days. And aha! A "moonth" is 28 days, where $28 = 1 + 2 + 4 + 7 + 14$. It's clear She *is* a mathematician! How convenient that a triangle with sides 3, 4, and 5 always has a right angle opposite its longest side. The triples $(5, 12, 13)$ and $(8, 15, 17)$ also yield a right-angled triangle. [In my umpiring days, I laid out softball diamonds using just such Pythagorean triples]. Is there a rule for finding *all* triples that provide a right-angle? [Yes. See below]. Are there lots of perfect numbers? [The next two are 496 and 8128.]

My first examples are trivial. The next is just numerology (*cf.* astrology). But the final questions nearly are real mathematics. Mind you, Euclid (~ -300) already knew the rule for finding all Pythagorean triples of integers (whole numbers). It's Diophantus's report of this result that sparks Fermat's infamous marginal note.

We know almost nothing about the man Diophantus except that he lived; in Alexandria. [Our only certain knowledge of Diophantus rests upon the fact that he quotes Hypsicles (~ -150) and that he is quoted by Theon Alexandrinus (whose date is fixed by the solar eclipse of June 16, 364)].

Diophantus's work was lost with the burning of the Library of Alexandria. More than a thousand years later, in 1464, six of its thirteen books were found in a library in Germany.

It is our great fortune that Claude Bachet's excellent translation into Latin came into the hands of Fermat in Toulouse. Number Theory was strong in antiquity. Surprisingly perhaps, when it came to inspire Fermat it was still a hundred years ahead of the then mathematics.

Diophantus presents a collection of problems and their solution. Typically, one is asked to find rational numbers (fractions) satisfying some equation in two unknowns. A random example, Problem 24 of Book VI, asks to split the number 6 into two parts, say y and $6 - y$, so that the product of those parts is the difference of a cube minus its cube root x. [The smallest nontrivial solution is $x = 17/9$, $y = 26/27$]. Fermat understands the principles behind Diophantus's methods. He realizes that the correct question is to ask for *all* rational solutions, not just for one.

Of course, Fermat doesn't *publish*. Rather, he boasts to his friends by challenging them to duplicate his solutions. Such challenges can be cruel. In 1643 Fermat writes to his friend Marin Mersenne asking him to find the smallest right triangle so that the hypotenuse c is a square and the sum $a + b$ of the other two sides also is a square. Innocent enough, one might think. But as Fermat could discover, using a generalization of the ideas of Diophantus, the smallest solution in whole numbers has $a = 1061652293520$ and $c = 4687298610289$ [Exercise for the reader: Coax your calculator to find b and check that this is a solution]. Poor Mersenne!

Fermat did leave us his proof showing that a fourth power is not the sum of two fourth powers. Both a challenge to Frénicle in 1638 and our understanding of his methods suggest that Fermat might also have been able to prove the Last Theorem for cubes. But the evidence is overwhelming that he quickly realized that he was mistaken in the matter of a general argument. He doesn't crow about the general theorem to his friends. He doesn't mention the "Last Theorem" in the final 25 years of his life. So when I'm asked to prove the Theorem the way Fermat did I say, "That's easy. One makes a foolish mistake and realizes it the next morning. But one forgets to rub out one's scribbled claim in the margin. And one forgets to tell one's son not to reprint one's marginal notes uncritically."

I nearly forgot to say a word about Pythagoras's Theorem. Problem 8 in Book II of *Diophantus* asks for a rule for writing a square as the sum of two squares. The resulting equation is that of the Theorem of Pythagoras, which says that in every right-angled triangle the square on the hypotenuse is the sum of the squares on the other two sides. The logo of Macquarie University's ceNTRe for Number Theory Research

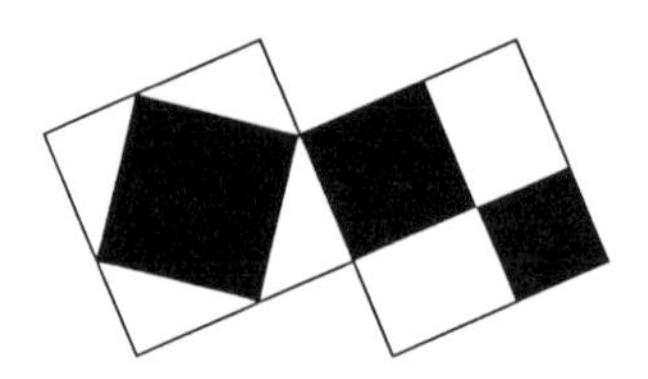

provides a graphical proof. The two larger squares each contain four identical right-angled triangles, and have sides of the same length, namely the sum of the lengths of the two shorter sides of the right-angled triangles. Thus the remaining areas in the two squares are equal. In the left-hand square, this area is the area of the square drawn on the hypotenuse of the right-angled triangle. In the right-hand square it is the sum of the areas of the squares drawn on the other two sides.

One obtains *all* integer solutions of

$$a^2 + b^2 = c^2$$

with b even, from

$$a = u^2 - v^2, \quad b = 2uv, \quad c = u^2 + v^2 .$$

It is redundant for a, b, and c to have a common factor. So one should choose u and v integers (whole numbers) with no common factor, such that one is odd and the other even.

From Fermat to 1993

By 1993, we could prove that no n th power — with n less than four million — is the sum of two n th powers. "Wow!", you say. "Surely that's enough." I suppose it is — for engineers. But compared to infinity, four million is just zilch.

Almost a century after Fermat, we see the true flowering of modern number theory. Euler is told by Goldbach about Fermat's remarks and is influenced by them; just as Diophantus inspired Fermat. Euler proves a great many things. Among them, by no means the highlight, that, indeed, a cube of an integer cannot be the sum of cubes of two nonzero integers. His method may well have been different from the one that Fermat possibly used.

Euler guesses, as have too many after him, that it all stands to reason. A square doesn't mind being split into *two* squares. But, of course, a cube wants at least *three* cubes. And no doubt, a fourth power refuses to split into fewer than *four* fourth powers, and so on.

So it was fun in the sixties — two hundred years later — when Lander and Parkin, putting a computer through its paces, came upon the fact that 61917364224 is the sum of 14348907, 4182119424, 16105100000, and 18424351793. [Exercise: Show that this displays a fifth power, namely of 144, as the sum of just *four* fifth powers.]

Just a few years ago, Noam Elkies at Harvard proved that there are infinitely many cases for which a fourth power *is* the sum of just *three* fourth powers. The *smallest* example reports that the fourth power of 422481 is the sum of the fourth powers, respectively, of 95800, 217519, and of 414560.

So even the best can stumble. Just so, Euler found that Fermat had stumbled in suggesting that the sequence

$$3, 5, 17, 257, 65\,537, 4\,294\,967\,297, \ldots$$

might consist just of prime numbers. It's easy to discover, by the very ideas developed by Fermat, that 641 divides 4 294 967 297. Apparently, Fermat got his sums wrong and never went back to check his calculations. [Exercise: What is this sequence of "Fermat Numbers"? Much sillier: Why, by the way, do we reckon that 9 is the next integer in the sequence 3, 1, 4, 1, 5, ...?]

Why is it, you ask, if Fermat could blunder and Euler could conjecture wrongly, that we can be so confident that cranks, claiming to prove the FLT, are to be dismissed. Your question, itself, is utterly illogical. You're thinking, Fermat made mistakes, cranks make mistakes. Cranks are as good as Fermat. Up a gum tree! Why not throw in the usual: Galileo was persecuted. Galileo was right. Cranks are derided ... .

The claims of cranks are characterized by historical errors, vagueness and lack of clarity, an unduly simple answer to a complex problem, confusing notation, grandiose claims for the significance of the result, secrecy about some key ideas, private publication, persecution mania, rejection of standard mathematics, and truisms presented as profundities; and I might add, an unreasonable reluctance to acknowledge any mistake.

Interest in Fermat's Last Theorem was revived in France in 1816 by the Parisian Academy offering a medal and a substantial prize. It had been many years after Fermat, in 1753, that Euler had dealt with the case of exponent $n = 3$ [that is, with cubes]. There was an alleged omission in the argument, later dealt with by Gauss. Dirichlet and Legendre proved the case $n = 5$ [fifth powers] in 1825 and Lamé settled the case $n = 7$ in 1839; Dirichlet had proved the case $n = 14$ in 1832.

In the meantime, Sophie Germain deals with many subcases. She shows that the "first case" of the FLT, when the exponent n does not divide any of the three integers involved, has no solution for many n. There was no equal opportunity in those days. Her work was presented to the Academy by Legendre.

The real breakthrough came from Germany. Applying ideas developed for a much more serious purpose — the higher reciprocity laws — Kummer seemed almost to settle Fermat's Last Theorem. Kummer shows that although the FLT is about honest-to-goodness whole numbers, it may be — perhaps, should be — viewed as a question about "complex" numbers built from ordinary integers and "imaginary" roots of unity. Kummer studies the arithmetic of these *cyclotomic* integers. He develops the notion of *ideal* divisor to cope with the fact that the familiar rule of unique factorization into primes fails for these "integers". And much more

Kummer shows that the arithmetic of cyclotomic fields behaves well enough to settle the FLT when the exponent is *regular*. He finds criteria for primes to be regular. The exceptions below 100 are 37, 59, and 67. Kummer then deals with these mildly irregular cases. He grinds to a halt at the somewhat more irregular case 167, when the calculations become just too burdensome.

Again in 1850 the Académie des Sciences de Paris offered a golden medal and a prize of 3000 francs to the mathematician who would solve Fermat's problem. In 1856 it determined that the question be withdrawn from competition but to instead award the medal to Kummer "for his beautiful researches on the complex numbers composed of roots of unity and integers".

While playing with Fermat's Last Theorem, Kummer had developed the subject now known as algebraic number theory. In his work one finds the germs of p-adic analysis. And .... I'll just say that it remains profitable to study Kummer. It's very respectable to have his "Collected Works" on one's shelves.

Even now, we still cannot *prove* that there are infinitely many regular primes. Sure, we "know" from experiment — extensive computation, and from heuristics — the feeling in our stomachs, that some 61% of primes are regular. But that's just physics, not mathematics.

In 1908 the substantial sum of 100, 000 Reichsmark was left by the will of Dr Paul Wolfskehl, "to be given to the person who will be the first to prove the great theorem of Fermat". In its first year the Wolfskehl Prize attracted 621 submissions! It's the romance, and size, of this prize that gave the FLT its popularity and notoriety. There have been no credible solutions submitted for the prize.

Still, let's pause to digest the meaning of the little we did know by 1993. Recent computation extended the range of Kummer's arguments to exponent 4 million, or so. It used work of the American mathematician Vandiver, from earlier this century; and of the Finn, Inkeri; among a large number of others. The sad fact remains though, that we still know too little about cyclotomic fields.

Suppose Fermat's remark were wrong. What would a counterexample look like? Work of Inkeri, and for that matter of mine, says that in a putative solution with exponent n the three numbers are themselves as large as n to the n th power. Whatever, if every particle in the universe had been counting since the Big Bang, the total sum reached by all would be as nothing compared to the numbers in that supposed solution. You can see that there could not be much practical importance to there being some accidental, monstrous solution.

The Last "Theorem" was surely just a scribble of Fermat's. But there is a perfectly *good* result of his, rather absurdly known as Fermat's "Little" Theorem. Its generalization by Euler is fundamental to much of modern cryptosystems. It might well be Fermat who helps stop the technician from just reading your PIN from the ATM.

Modern encryption relies on multiplication being a "trapdoor". It's not all that hard to multiply. Most of us could multiply two 20-digit numbers, given a sheet of paper and twenty minutes. But asked to "undo" our work, to *factorize* a 40-digit number ... ? A reasonable computer will multiply two 100-digit numbers in the blink of an electron, so to speak. It still takes a world-wide net of computers, and months of real time, to factor a "random" number of just 130 digits.

I had a period of serious interest in Fermat's Last Theorem in the seventies, partly because of some related work I had done which led to my writing a joint paper on the FLT with Kustaa Inkeri. Curiously, I also did some calculations on the Taniyama-Weil Conjecture at about that time, jointly with M. K. Agrawal, John Coates, and David Hunt (at UNSW), of course with no idea that it might be related. I was surprised to find how little of great substance seemed to have been achieved on the FLT in the previous fifty years. The eighties were to be different.

Fermat's claim says that if $n > 2$,

$$a^n + b^n = c^n$$

is impossible in integers a, b, c all nonzero.
According to Darmon and Granville (1993),

> Fermat might have chosen a different generalization. Led by this, I propose that if a, b, c are relatively prime, then
>
> $$a^t + b^u = c^v$$
>
> has no solution in integers greater than 1 if all of t, u, v are at least 3. If one exponent is allowed to be 2, things are different. For example, in the cases $(t, u, v) = (3, 3, 2)$ and $(4, 3, 2)$ there are infinitely many solutions. In general, if
>
> $$\frac{1}{t} + \frac{1}{u} + \frac{1}{v} < 1$$
>
> we have grounds for believing that there are just finitely many solutions for which a and b have no common factor. One of the large known solutions is
>
> $$43^8 + 96\,222^3 = 30\,042\,907^2 .$$
>
> All ten known solutions have one or other of the exponents equal to 2.

Now let me add that I feel a damned fool for not daring to write

$$144^5 = 27^5 + 84^5 + 110^5 + 133^5$$

and

$$422481^4 = 95800^4 + 217519^4 + 414560^4 .$$

But one feels constrained by the conviction that newspapers cannot print superscripts (or believe that to do so would frighten readers away).

From Diophantus to Wiles

Just what did Fermat read in Diophantus's *Arithmetica*? And what have we learned in the intervening 350 years? Diophantus's problems boil down to equations in two unknowns. The graphs of such equations are plane curves. Diophantus* wants solutions in rational numbers (fractions).

These problems fall into three classes. First there are "rational curves": these include the cases where the curve is of degree 1 or 2. Rational curves

*When these remarks first appeared, I had a complaint about my writing 'Diophantus' rather than 'Diophantos'. Eventually, I replied: "This seems a Cebyshev matter to me. When W. E. (Bill) Smith first introduced me to Tschebysheff polynomials he pointed out that one of the charms of the name 'Cebycev' was that one could spell it virtually any way one liked without necessarily being wrong. Tschebotarev shares this happy property."

have no rational points, or infinitely many. If there are infinitely many, they're given by a simple formula involving just "rational functions".

> All the rational points on the rational curve $x^2 + y^2 = 1$ are given by
>
> $$x = \frac{1 - t^2}{1 + t^2}, \quad y = \frac{2t}{1 + t^2};$$
>
> with t any fraction.
>
> But there are no rational points on the curve $x^2 + y^2 = -1$.

At the other extreme are curves of "general type" (technically, of genus 2 or more). These have at most a few rational solutions. There's little rhyme or pattern. The interesting case is the case of "elliptic curves", typified by polynomial equations of degree 3. Here Diophantus suggests and Fermat develops a clever technique, "the chord and tangent method", for obtaining new solutions from old ones.

The *congruent number* problem, dating back to an Arabic manuscript of 932, asks which integers n are the common difference in an arithmetic progression of three rationals squared. For example, because of 1, 25, 49 we see that $24 = 2^2 \times 6$ (and hence 6) is such a *congruent* number. Fermat could show that there cannot be an arithmetic progression of four rationals squared. In 1225 (itself a square!) Leonardo of Pisa (known as Fibonacci) showed that 5 is congruent because of the sequence 961/144, 1681/144, 2401/144.

It turns out [Exercise: This is not obvious; but not impossibly hard] that n is congruent if, and only if, it is the area of a right-angled triangle with rational sides. The triangle confirming that 5 is congruent has sides $(3/2, 20/3, 41/6)$. Nowadays, we can decide in practice whether an integer is congruent without knowing the sequence or the triangle. That's lucky. We know, for example, that 157 is congruent, but the confirming triangle has sides whose denominators are some 100 digits long.

A different problem, but in this spirit, is to determine those whole numbers which can be written as the sum (or difference) of cubes of fractions. That 7 is the sum of the cubes of 2 and of -1 is not very exciting. That 6 is the sum of the cubes of 17/21 and 37/21 is less obvious. Our methods nowadays can prove that, say, 382 must be the sum of two cubes. It's a good thing we can determine this fact by theory, because the actual smallest solution involves fractions with denominator

81220543934857938931677195009290600931518540131945 74

rather longer than the width of a newspaper column.

No one, certainly no sane mathematician, thinks it important that 157 is congruent; or that 382 is the sum of two rational cubes. Mathematics is concerned with understanding the situation. Mathematics is the developing of general theory that might be used to answer these and a multitude of related questions. Mathematics is concerned with inner truth, not with stupid puzzles, or even clever ones. But it's fun — childish really; but fun — to compute some particular dramatic cases. Even then the thing is that it can be done, not the having done it.

What then makes a good question? It's mostly a matter of whether the answer is interesting; I mean the methods used to produce the answer. If it's a really old question, I suppose we give it a grudging respect. It's a bit like being polite to our seniors.

Elliptic curves are interesting because they are the case for which it can be damnably subtle to decide whether they have infinitely many rational points or just a few. And they are the simplest case of a yet more profound structure, the *abelian variety*.

Is Fermat's Last Theorem important? Well, it's pretty old. It has provoked some interesting mathematics. But frankly, I doubt if mathematics would have been the poorer for son Samuel not reprinting that note in Fermat's copy of *Diophantus*.

Fermat's is actually lots of equations. From

$$a^n + b^n = c^n$$

on setting $x = a/c$ and $y = b/c$, we get

$$x^n + y^n = 1\,.$$

A solution like $x = 1$, $y = 0$ is trivial. When $n = 2$ we obtain a rational curve. The cases

$$x^3 + y^3 = 1 \quad \text{and} \quad x^4 + 1 = w^2$$

are elliptic curves with no nontrivial rational points (when $n = 4$ the trick was to set $x = a/b$ and $w = c^2/b^2$). When n is 5 or more we get curves of general type.

In 1983 Gerd Faltings proved Mordell's Conjecture that curves of genus at least 2 have only finitely many rational points. This required powerful methods from algebraic geometry, at first comprehensible only to experts.

So then we knew that each Fermat equation has, at worst, only a few solutions. But the methods introduced by Kummer told us that there are no nontrivial solutions at all for lots of n. We knew no good reason for that. A possible explanation came from a suggestion by Gerhard Frey.

Frey points out that the equation

$$y^2 = x(x - a^n)(x + b^n)$$

> is the equation of an elliptic curve. But *if* $a^n + b^n = c^n$ in integers a, b, and c all nonzero, then the Frey curve has properties that an elliptic curve shouldn't have. That's *supposing* the truth of the conjecture of Taniyama-Weil-Shimura. So all we need is to *prove* TWS and then we'll have proved the FLT.

In 1986 Frey [pronounced as in the English "fry", by the way] suggested we study an elliptic curve concocted from a supposed solution of Fermat's equation. A little later, Ken Ribet (Berkeley) *proved* that, indeed, this ain't on. Surely. If there were such a Frey curve, then a very fundamental conjecture about elliptic curves (and about lots of other things) would be false.

Up to this moment, we didn't think that the FLT was a matter of inner truth. We barely cared whether there was some sporadic solution or not. We didn't think that nature cared. But, by crikey, we care about the Taniyama-Shimura-Weil Conjecture. If it were false, then inner truth is very different from what we've been thinking. The trouble is, until the other day we feel that our *proving* TWS is still a generation away.

Incidentally. Why *elliptic* curve, you may be asking. Are these ellipses or something? No. Ellipses are just rational curves. But just as trigonometric functions [mathematicians speak of *circular functions*] crop up in looking at the arc length of circles, so new functions — naturally enough called *elliptic functions* — occur in studying arc length of ellipses. Those functions happen to parametrize elliptic curves.

What is this TWS Conjecture all about? Well, it suggests that elliptic curves defined over the rationals have a yet richer structure than first hits the eye. They're also controlled by certain automorphic functions. If that's really so, then we can explain lots of uncoordinated facts we've been noticing in the last 350 years.

Frey told me the other day (there was a Fermat meeting at the Chinese University, Hong Kong) about mentioning his thought, as little more than a passing joke, at the Séminaire de Théorie de Nombres de Paris in 1986. He described Andrew Wiles rising to complain. This *would* be the way that Fermat's Last Theorem will be proved, Wiles declared passionately.

The legend will go on to say that most of us didn't think so; or that we didn't care. But Wiles went off to prove enough of the TWS to entail the truth of the FLT. And here he has astonished us. Not because he may have proved the FLT. Who cares? Though it'll be nice to give Wiles the Wolfskehl Prize and to get the cranks off our backs.

The thing is, Wiles has proved large portions of the TWS Conjecture. That's for sure. We now know new ideas suggesting the proof of the full conjecture may be just months away. Surely we will not have to wait a generation. We've had a blinding glimpse of inner truth.

The study of elliptic curves still is dominated by various conjectures. Given the conjectures, all sorts of computations are possible, and their success provides compelling evidence in support of the conjectures. Were we talking physics, these "theories" would be deemed to be quite well established. But we're talking about the ideal world, not just the real one. As mathematicians all we can do is to pat our stomachs, or whatever else we deem to be the site of our beliefs, and reconfirm our "feelings". It is in this context that Wiles' result is so important. One can never be certain about one's stomach; it may only be wind. Wiles confirms that certain views that have driven our investigations are *true* and are not just some astonishingly effective approximation that happens to work for the relatively small numbers accessible to us and our computers.

But the whole thing is muddied by that stupid Fermat problem. At this moment Wiles' arguments do not quite prove that Frey curves can't exist. So silly people say that Wiles has failed. I'll gladly volunteer to "fail" his way!

Do you want to know about the "gap"? OK. Wiles remembers that the L-function of the symmetric square of an elliptic curve does not vanish at the critical point. So all's well. But Wiles sort of forgets that the p-adic analogues of the L-function always have a zero. It shouldn't matter. But that kind of creates a quasi-zero: it's called a *trivial zero* in the trade. That's bad. Satisfied?

We're not sure that Wiles has gotten around the problem. It barely affects his work. But it does affect the tiny extra bit that happens to be needed for the FLT.

But I've gotten too serious. So let me tell you a fun thing. Now that it seems that we might be able to prove TWS there's almost a brawl about who should be credited with the conjecture. There are arguments for including, or for omitting, any of the names of Taniyama, Shimura, or Weil. In what order should they be mentioned? In Hong Kong speakers spoke mysteriously about "a *celebrated* conjecture". Serre suggested we call it the $***$-Conjecture. It seems likely that the anonymous name "The Modularity Conjecture" will win out. Better yet, and happily it sounds just like the old "Taniyama-Weil Conjecture", the "Taniyama-Wiles Theorem".

I hope you enjoyed the glimpse of mathematics I may have shown you. Not to worry. Fermat was right. His equation has no nontrivial solutions once n is bigger than 2. Kummer sidetracked us a bit. But we now know what the proof will look like. Wiles has shown us what to do. It'll be finished one day. Probably quite soon now. Our understanding of inner truth will not be much the different for it.

[On October 25, 1994 Wiles released a pair of papers, his 'Modular elliptic curves and Fermat's Last Theorem' and a joint addendum with Richard

Taylor, 'Ring-theoretic properties of certain Hecke algebras', which settle the matter after all.

It seems that Matthias Flach's "Euler systems", which Wiles had thought to provide the extra to justify his 1993 announcement, cannot as yet be made to work in the present context. But by returning to his "Plan A", Wiles has found that he could prove Fermat's Last Theorem, much as had been announced.

Let me conclude by reiterating the Generalized Fermat Conjecture. This GFC alleges that the equation $x^t + y^u = z^v$ has no solution, in nonzero integers (x, y, z), that do not share a common factor, if the exponents (t, u, v) all are greater than 2. We believe, there are just finitely many solutions anyhow if $1/t + 1/u + 1/v < 1$. In the past year Frits Beukers (Utrecht) has shown that in the contrary case there are indeed infinitely many solutions but all belong to just finitely many parametrized families. Henri Darmon, in part responsible for inspiring the GFC, is, incidentally, credited by Wiles for having questioned him at length on the matter of Wiles' Plan A, thus leading Wiles to realize that his original approach, without the Euler systems, was indeed capable of success].

Andrew Wiles

Photograph by C. J. Mozzochi, Princeton

get $h_\infty h_p \leq 1$. Then with this hypothesis, $(W^0_{\lambda^n})^*$ is easily verified to be unramified with Frob p acting as $U_p^2\langle p\rangle^{-1}$ by the description of $\rho_{f,\lambda}|_{D_p}$ in [Wil, Th. 2.1.4].) On the other hand, we have constructed classes which are ramified at primes in Q in (3.7). These are of type $\mathcal{D}_Q$. We also have classes in

$$\mathrm{Hom}(\mathrm{Gal}(\mathbf{Q}_{\Sigma\cup Q}/\mathbf{Q}), \mathcal{O}/\lambda^M) = H^1(\mathbf{Q}_{\Sigma\cup Q}/\mathbf{Q}, \mathcal{O}/\lambda^M) \hookrightarrow H^1(\mathbf{Q}_{\Sigma\cup Q}/\mathbf{Q}, V_{\lambda^M})$$

coming from the cyclotomic extension $\mathbf{Q}(\zeta_{q_1}\ldots\zeta_{q_r})$. These are of type $\mathcal{D}$ and disjoint from the classes obtained from (3.7). Combining these with (3.10) gives

$$\# H^1_{\mathcal{D}}(\mathbf{Q}_\Sigma/\mathbf{Q}, V_f[\lambda^M]) \leq t \cdot \#(\mathfrak{p}_{\mathbf{T}}/\mathfrak{p}^2_{\mathbf{T}}) \cdot c_p$$

as required. This proves part (i) of Theorem 3.1.

Now if we assume that $\mathbf{T}$ is a complete intersection we have that $t = 1$ by Proposition 2 of the appendix. In the strict or flat cases (and indeed in all cases where $c_p = 1$) this implies that $R_{\mathcal{D}} \simeq \mathbf{T}_{\mathcal{D}}$ by Proposition 1 of the appendix together with Proposition 1.2. In the Selmer case we get

$$\text{(3.11)} \qquad \#(\mathfrak{p}_{\mathbf{T}}/\mathfrak{p}^2_{\mathbf{T}}) \cdot c_p = \#(\mathcal{O}/\eta_{\mathbf{T},f})c_p = \#(\mathcal{O}/\eta_{\mathbf{T}_{\mathcal{D}},f}) \leq \#(\mathfrak{p}_{\mathbf{T}_{\mathcal{D}}}/\mathfrak{p}^2_{\mathbf{T}_{\mathcal{D}}})$$

where the central equality is by Remark 2.18 and the right-hand inequality is from the theory of Fitting ideals. Now applying part (i) we see that the inequality in (3.11) is an equality. By Proposition 2 of the appendix, $\mathbf{T}_{\mathcal{D}}$ is also a complete intersection.

The final assertion of the theorem is proved in exactly the same way on noting that we only used the minimality to ensure that the h_q's were 1. In general, they are bounded independent of M and easily computed. (The only point to note is that if $\rho_{f,\lambda}$ is of multiplicative type at q then $\rho_{f,\lambda}|_{D_q}$ does not split.) □

Remark. The ring $\mathbf{T}_{\mathcal{D}_0}$ defined in (3.1) and used in this chapter should be the deformation ring associated to the following deformation problem $\mathcal{D}_0$. One alters $\mathcal{D}$ only by replacing the Selmer condition by the condition that the deformations be flat in the sense of Chapter 1, i.e., that each deformation ρ of ρ_0 to $\mathrm{GL}_2(A)$ has the property that for any quotient $A/\mathfrak{a}$ of finite order, $\rho|_{D_p} \bmod \mathfrak{a}$ is the Galois representation associated to the $\bar{\mathbf{Q}}_p$-points of a finite flat group scheme over $\mathbf{Z}_p$. (Of course, ρ_0 is ordinary here in contrast to our usual assumption for flat deformations.)

From Theorem 3.1 we deduce our main results about representations by using the main result of [TW], which proves the hypothesis of Theorem 3.1 (ii), and then applying Theorem 2.17. More precisely, the main result of [TW] shows that $\mathbf{T}$ is a complete intersection and hence that $t = 1$ as explained above. The hypothesis of Theorem 2.17 is then given by Theorem 3.1(i), together with the equality $t = 1$ (and the central equality of (3.11) in the

I'm starting to see the attraction,
in deformation 'n complete intersection.
I no longer fear 'em,
'cause Fermat's Last Theorem
Demands that they have our affection.

Annals of Mathematics **141** (3), May 1995

Appendix B

"The Devil and Simon Flagg"

by Arthur Porges

After several months of the most arduous research, involving the study of countless faded manuscripts, Simon Flagg succeeded in summoning the devil. As a competent medievalist, his wife had proved invaluable. A mere mathematician himself, he was hardly equipped to decipher Latin holographs, particularly when complicated by rare terms from tenth-century demonology, so it was fortunate that she had a flair for such documents.

The preliminary skirmishing over, Simon and the devil settled down to bargain in earnest. The devil was sulky, for Simon had scornfully declined several of his most dependable gambits, easily spotting the deadly barb concealed in each tempting bait.

"Suppose you listen to a proposition from me for a change," Simon suggested finally. "At least, it's a straightforward one."

The devil irritably twisted his tail-tip with one hand, much as a man might toy with his key chain. Obviously, he felt injured.

"All right," he agreed, in a grumpy voice. "It can't do any harm. Let's hear your proposal."

"I will pose a certain question." Simon began, and the devil brightened, "to be answered within twenty-four hours. If you cannot do so, you must pay me $100,000. That's a modest request compared to most you get. No billions, no Helen of Troy on a tiger skin. Naturally there must be no reprisals of any kind if you win."

"Indeed!" the devil snorted. "And what are your stakes?"

"If I lose, I will be your slave for any short period. No torment, no loss of soul—not for a mere $100,000. Neither will I harm relatives or friends." Although, he amended thoughtfully, "there are exceptions."

The devil scowled, pulling his forked tail petulantly. Finally, a savage tug having brought a grimace of pain, he desisted.

"Sorry," he said flatly. "I deal only in souls. There is no shortage of slaves. The amount of free, wholehearted service I receive from humans

---

would amaze you. However, here's what I'll do. If I can't answer your question in the given time, you will receive not a paltry $100,000, but any sum within reason. In addition, I offer health and happiness as long as you live. If I do answer it—well, you know the consequences. That's the very best I can offer." He pulled a lighted cigar from the air and puffed in watchful silence.

Simon stared without seeing. Little moist patches sprang out upon his forehead. Deep in his heart he had known what the devil's only terms would be. Then his jaw muscles knotted. He would stake his soul that nobody—man, beast, or devil—could answer this question in twenty-four hours.

"Include my wife in that health and happiness provision, and it's a deal," he said. "Let's get on with it."

The devil nodded. He removed the cigar stub from his mouth, eyed it distastefully, and touched it with a taloned forefinger. Instantly it became a moist pink mint, which he sucked with noisy relish.

"About your question," he said, "it must have an answer, or our contract becomes void. In the Middle Ages, people were fond of proposing riddles. A few came to me with paradoxes, such as that one about a barber who shaves all those, and only those, who don't shave themselves. 'Who shaves the barber?' they asked. Now, as Russell has noted, the 'all' makes such a question meaningless and so unanswerable."

"My question is just that—not a paradox," Simon assured him.

"Very well. I'll answer it. What are you smirking about?"

"Nothing," Simon replied, composing his face.

"You have very good nerves," the devil said, grimly approving, as he pulled a parchment from the air. "If I had chosen to appear as a certain monster which combines the best features of your gorilla with those of the Venusian Greater Kleep, an animal—I suppose one could call it that—of unique eye appeal, I wonder if your aplomb . . . "

"You needn't make any tests," Simon said hastily. He took the proffered contract and, satisfied that all was in order, opened his pocketknife.

"Just a moment," the devil protested. "Let me sterilize that; you might get infected." He held the blade to his lips, blew gently, and the steel glowed cherry red. "There you are. Now a touch of the point to some—ah—ink, and we're all set. Second line from the bottom, please; the last one's mine."

Simon hesitated, staring at the moist red tip.

"Sign," urged the devil, and, squaring his shoulders, Simon did so. When his own signature had been added with a flourish, the devil rubbed his palms together, gave Simon a frankly proprietary glance, and said jovially, "Let's have the question. As soon as I answer it, we'll hurry off. I've just time for another client tonight."

"All right," said Simon. He took a deep breath. "My question is this: Is Fermat's Last Theorem correct?"

The devil gulped. For the first time his air of assurance weakened.

"Whose last what?" he asked in a hollow voice.

"Fermat's Last Theorem. It's a mathematical proposition which Fermat, a seventeenth-century French mathematician, claimed to have proved. However, his proof was never written down, and to this day nobody knows if the Theorem is true or false." His lips twitched briefly as he saw the devil's expression. "Well, there you are—go to it!"

"Mathematics!" the devil exclaimed, horrified. "Do you think I've had time to waste learning such stuff? I've studied the Trivium and Quadrivium, but as for algebra—say," he added resentfully, "what kind of question is that to ask me?"

Simon's face was strangely wooden, but his eyes shone. "You'd rather run 75,000 miles and bring back some object the size of Boulder Dam, I suppose!" he jeered. "Time and space are easy for you, aren't they? Well, sorry. I prefer this. It's a simple matter," he added, in a bland voice. "Just a question of positive integers."

"What's a positive integer?" the devil flared. " ... or an integer, for that matter?"

"To put it more formally," Simon said, ignoring the devil's question, "Fermat's Last Theorem states that there are no non-trivial, rational solutions of the equation $x^n + y^n = z^n$, for n a positive integer greater than two."

"What's the meaning of— "

"You supply the answers, remember."

"And who's to judge—you?"

"No," Simon replied sweetly. "I doubt if I'm qualified, even after studying the problem for years. If you come up with a solution, we'll submit it to any good mathematical journal, and their referee will decide. And you can't back out—the problem obviously is soluble: either the theorem is true, or is it false. No nonsense about multi-valued logic, mind. Merely determine which, and prove it in twenty-four hours. After all, a man—excuse me—demon, of your intelligence and vast experience surely can pick up a little math in that time."

"I remember now what a bad time I had with Euclid when I studied at Cambridge," the devil said sadly. "My proofs were always wrong, and yet it was all obvious anyway. You could see just by the diagrams." He set his jaw. "But I can do it. I've done harder things before. Once I went to a distant star and brought back a quart of neutronium in just sixteen— "

"I know," Simon broke in. "You're very good at such tricks."

"Trick, nothing!" was the angry retort. "It's a technique so difficult—but never mind, I'm off to the library. By this time tomorrow— "

"No," Simon corrected him. "We signed half an hour ago. Be back in exactly twenty-three point five hours! Don't let me rush you," he added ironically, as the devil gave the clock a startled glance. "Have a drink and meet my wife before you go."

"I never drink on duty. Nor have I time to make the acquaintance of your wife ... now." He vanished.

The moment he left, Simon's wife entered.

"Listening at the door again?" Simon chided her, without resentment.

"Naturally," she said in a throaty voice. "And darling—I want to know—that question—is it really difficult? Because if it's not—Simon, I'm so worried."

"It's difficult, all right." Simon was almost jaunty. "But most people don't realize at first. You see," he went on, falling automatically into his stance for Senior Math II, "anybody can find two whole numbers whose squares add up to a square. For example, $3^2 + 4^2 = 5^2$; that is, $9 + 16 = 25$. See?"

"Uh huh." She adjusted his tie.

"But when you try to find two cubes that add up to a cube, or higher powers that work similarly, there don't seem to be any. Yet," he concluded dramatically, "nobody has been able to prove that no such numbers exist. Understand now?"

"Of course." Simon's wife always understood mathematical statements, however abstruse. Otherwise, the explanation was repeated until she did, which left little time for other activities.

"I'll make us some coffee," she said, and escaped.

Four hours later as they sat together listening to Brahms' Third, the devil reappeared.

"I've already learned the fundamentals of algebra, trigonometry, and plane geometry!" he announced triumphantly.

"Quick work," Simon complimented him. "I'm sure you'll have no trouble at all with spherical, analytic, projective, descriptive, and non-Euclidean geometries."

The devil winced. "Are there so many?" he inquired in a small voice.

"Oh, those are only a few." Simon had the cheerful air suited to a bearer of welcome tidings. "You'll like non-Euclidean," he said mendaciously. "There you don't have to worry about diagrams—they don't tell a thing! And since you hated Euclid anyway— "

With a groan the devil faded out like an old movie. Simon's wife giggled.

"Darling," she sang, "I'm beginning to think you've got him over a barrel."

"Ssh," said Simon. "The last movement. Glorious!"

Six hours later, there was a smoky flash, and the devil was back. Simon noted the growing bags under his eyes. He suppressed a grin.

"I've learned all those geometries," the devil said with grim satisfaction. "It's coming easier now. I'm about ready for your little puzzle."

Simon shook his head. "You're trying to go too fast. Apparently you've overlooked such basic techniques as calculus, differential equations, and finite differences. Then there's— "

"Will I need all those?" the devil moaned. He sat down and knuckled his puffy eyelids, smothering a yawn.

"I couldn't say," Simon replied, his voice expressionless. "But people have tried practically every kind of math there is on that 'little puzzle', and it's still unsolved. Now, I suggest— " But the devil was in no mood for advice from Simon. This time he even made a sloppy disappearance while sitting down.

"I think he's tired," Mrs Flagg said. "Poor devil." There was no discernible sympathy in her tones.

"So am I," said Simon. "Let's go to bed. He won't be back until tomorrow, I imagine."

"Maybe not," she agreed, adding demurely, "but I'll wear the black lace—just in case."

It was the following afternoon. Bach seemed appropriate somehow, so they had Landowska on.

"Ten more minutes," Simon said. "If he's not back with a solution by then, we've won. I'll give him credit; he could get a Ph.D. out of my school in one day—with honors! However— "

There was a hiss. Rosy clouds mushroomed sulphurously. The devil stood before them, steaming noisomely on the rug. His shoulders sagged; his eyes were bloodshot; and a taloned paw, still clutching a sheaf of papers, shook violently from fatigue or nerves.

Silently, with a kind of seething dignity, he flung the papers to the floor, where he trampled them viciously with his cloven hoofs. Gradually then, his tense figure relaxed, and a wry smile twisted his mouth.

"You win, Simon," he said, almost in a whisper, eying him with ungrudging respect. "Not even I can learn enough mathematics in such a short time for so difficult a problem. The more I got into it, the worse it became. Non-unique factoring, ideals— Baal! Do you know," he confided, "not even the best mathematicians on other planets—all far ahead of yours—have solved it? Why, there's a chap on Saturn—he looks something like a mushroom on stilts—who solves partial differential equations mentally; and even he's given up." The devil sighed. "Farewell." He dislimned with a kind of weary precision.

Simon kissed his wife—hard. A long while later she stirred in his arms.

"Darling," she pouted, peering into his abstracted face, "what's wrong now?"

"Nothing—except I'd like to see his work; to know how close he came. I've wrestled with that problem for— " He broke off amazed as the devil flashed back. Satan seemed oddly embarrassed.

"I forgot," he mumbled. "I need to—ah!" He stooped for the scattered papers, gathering and smoothing them tenderly. "It certainly gets you," he said, avoiding Simon's gaze. "Impossible to stop just now. Why, if I could only prove one simple little lemma— " He saw the blazing interest in Simon, and dropped his apologetic air. "Say," he grunted, "you've worked on this, I'm sure. Did you try continued fractions? Fermat must have used them, and—move over a minute, please— " This last to Mrs Flagg. He sat down beside Simon, tucked his tail under, and pointed to a jungle of symbols.

Mrs Flagg sighed. Suddenly the devil seemed a familiar figure, little different from old Professor Atkins, her husband's colleague at the university. Any time two mathematicians got together on a tantalizing problem .... Resignedly she left the room, coffeepot in hand. There was certainly a long session in sight. She knew. After all, she was a professor's wife.

But it did take him more than a day ...

Photograph by C. J. Mozzochi, Princeton

Appendix C

"Math Riots"
by EricZorn

The following feature from the *Chicago Tribune* of June 29, 1993 is a dramatic exception to the rule that the inanity of newspaper articles on mathematical topics should invariably infuriate mathematician readers. Non-denizens of North America may need to be reminded of celebrations of victories of the Chicago Bulls and of the Chicago Blackhawks, parodied below. [Mind you, I'm not entirely clear that the settling of the FLT should decently be compared, say, to St George winning a premiership.] The allusions to Miyaoka refer unkindly to the previous occasion that a probable proof of the FLT had been announced. In effect, Miyaoka had thought himself able to move certain arguments from the function field to the arithmetic case, thus proving a version of the ABC-Conjecture and, in particular, Fermat's Last Theorem.

Math Riots Prove Fun Incalculable

by **Eric Zorn**

News Item (June 23) — *Mathematicians worldwide were excited and pleased today by the announcement that Princeton University professor Andrew Wiles had finally proved Fermat's Last Theorem, a 365-year-old problem said to be the most famous in the field.*

Yes, admittedly, there was rioting and vandalism last week during the celebration. A few bookstores had windows smashed and shelves stripped, and vacant lots glowed with burning piles of old dissertations. But overall we can feel relief that it was nothing — nothing — compared to the outbreak of exuberant thuggery that occurred in 1984 after Louis DeBranges finally proved the Bieberbach Conjecture.

"Math hooligans are the worst," said a Chicago Police Department spokesman. "But the city learned from the Bieberbach riots. We were ready for them this time."

'Math Riots', by Eric Zorn, appeared for a moment on the net, in breach of copyright, following its publication in the *Chicago Tribune* on June 29, 1993.

When word hit Wednesday that Fermat's Last Theorem had fallen, a massive show of force from law enforcement at universities all around the country headed off a repeat of the festive looting sprees that have become the traditional accompaniment to triumphant breakthroughs in higher mathematics.

Mounted police throughout Hyde Park kept crowds of delirious wizards at the University of Chicago from tipping over cars on the midway as they first did in 1976 when Wolfgang Haken and Kenneth Appel cracked the long-vexing Four-Color Problem. Incidents of textbook-throwing and citizens being pulled from their cars and humiliated with difficult mathematical problems last week were described by the university's math department chairman Bob Zimmer as "isolated".

Zimmer said, "Most of the celebrations were orderly and peaceful. But there will always be a few — usually graduate students — who use any excuse to cause trouble and steal. These are not true fans of Andrew Wiles."

Wiles himself pleaded for calm even as he offered up the proof that there is no solution to the equation $x^n + y^n = z^n$ when n is a whole number greater than two, as Pierre de Fermat first proposed in the 17th Century. "Party hard but party safe," he said, echoing the phrase he had repeated often in interviews with scholarly journals as he came closer and closer to completing his proof.

Some authorities tried to blame the disorder on the provocative taunting of Japanese mathematician Yoichi Miyaoka. Miyaoka thought he had proved Fermat's Last Theorem in 1988, but his claims did not bear up under the scrutiny of professional referees, leading some to suspect that the fix was in. And ever since, as Wiles chipped away steadily at the Fermat problem, Miyaoka scoffed that there would be no reason to board up windows near universities any time soon; that God wanted Miyaoka to prove it.

In a peculiar sidelight, Miyaoka recently took the trouble to secure a U.S. trademark on the equation "$x^n + y^n = z^n$" as well as the now-ubiquitous expression "Take that, Fermat!" Ironically, in defeat, he stands to make a good deal of money on cap and T-shirt sales.

This was no walk-in-the-park proof for Wiles. He was dogged, in the early going, by sniping publicity that claimed he was seen puttering late one night doing set theory in a New Jersey library when he either should have been sleeping, critics said, or focusing on arithmetic algebraic geometry for the proving work ahead.

"Set theory is my hobby, it helps me relax," was his angry explanation. The next night, he channeled his fury and came up with five critical steps in his proof. Not a record, but close.

There was talk that he thought he could do it all by himself, especially when he candidly referred to University of California mathematician Kenneth Ribet as part of his "supporting cast," when most people in the field knew that without Ribet's 1986 proof definitively linking the Taniyama Conjecture to Fermat's Last Theorem, Wiles would be just another frustrated guy in a tweed jacket teaching calculus to freshmen.

His travails made the ultimate victory that much more explosive for math buffs. When the news arrived, many were already wired from caffeine consumed at daily colloquial teas, and they took to the streets en masse shouting, "Obvious! Yessss! It was obvious!"

The law cannot hope to stop such enthusiasm, only to control it. Still, one has to wonder what the connection is between wanton pillaging and a mathematical proof, no matter how long-awaited and subtle.

The Victory Over Fermat rally, held on a cloudless day in front of a crowd of 30,000 (police estimate: 150,000) was pleasantly peaceful. Signs unfurled in the audience proclaimed Wiles the greatest mathematician of all time, though partisans of Euclid, Descartes, Newton, and C.F. Gauss and others argued the point vehemently.

A warmup act, The Supertheorists, delighted the crowd with a ragged song, "It Was Never Less Than Probable, My Friend", which included such gloating, barbed verses as—

I had a proof all ready
But then I did a choke-a
Made liberal assumptions
Hi! I'm Yoichi Miyaoka.

In the speeches from the stage, there was talk of a dynasty, specifically that next year Wiles will crack the great unproven Riemann Hypothesis ("Rie-peat! Rie-peat!" the crowd cried), and then after that the Prime Pair Problem, the Goldbach Conjecture ("Minimum Goldbach", said one T-shirt) and so on.

They couldn't just let him enjoy his proof. Not even for one day. Math people. Go figure 'em.

Fermat today:
Lycée Pierre de Fermat, Toulouse

Photograph by Kate van der Poorten

Index